RUNNING OUT

running out

in search of water on the high plains

lucas bessire

PRINCETON UNIVERSITY PRESS

PRINCETON AND OXFORD

Princeton University Press is committed to the protection of copyright
and the intellectual property our authors entrust to us. Copyright
promotes the progress and integrity of knowledge. Thank you for
supporting free speech and the global exchange of ideas by purchasing
an authorized edition of this book. If you wish to reproduce or distribute
any part of it in any form, please obtain permission.

Requests for permission to reproduce material from this work
should be sent to permissions@press.princeton.edu

Published by Princeton University Press
41 William Street, Princeton, New Jersey 08540
6 Oxford Street, Woodstock, Oxfordshire OX20 1TR

press.princeton.edu

All Rights Reserved
ISBN 978-0-691-21264-7
ISBN (e-book) 978-0-691-21265-4

British Library Cataloging-in-Publication Data is available

Editorial: Fred Appel and James Collier
Production Editorial: Jenny Wolkowicki
Text design: Karl Spurzem
Jacket design: Daniel Benneworth-Gray
Production: Erin Suydam
Publicity: Maria Whelan and Kathryn Stevens
Copyeditor: Maia Vaswani

Jacket credit: Shutterstock

This book has been composed in Arno

Printed on acid-free paper. ∞

Printed in the United States of America

10 9 8 7 6 5 4 3 2 1

For Tony and Lila Fern

CONTENTS

Map of the Ogallala Aquifer, 2015   viii

Map of the Little Rock House   ix

Note to the Reader   xi

Lines   1

Bones   39

Dust   93

Clouds   135

Afterword   167

Acknowledgments   183

Notes   187

Bibliography   209

List of Illustrations   237

Index   239

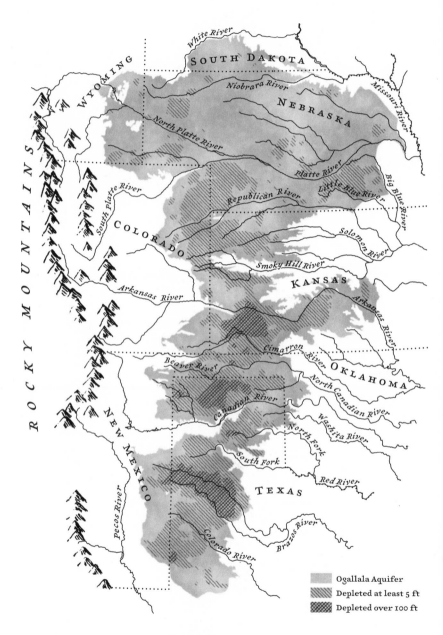

The Ogallala Aquifer, 2015

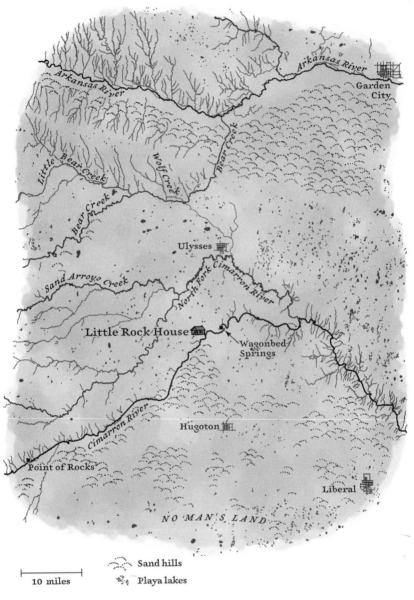

The Little Rock House

NOTE TO THE READER

The Ogallala aquifer underlies a vast expanse of America's heartlands. It was created by ancient seas buried when the western mountains began to rise at the dawn of time. Under the waters are the bedrocks of genesis. Over them sediments from a million years of wind mingle with the elemental forms of lives past and future. The deep waters are the fulcrum, the mystery, the source. Within them are traced all possibilities.

For ages the waters lay covered by earth. They rested, deep and cold, beyond human design. It took less than eighty years for settlers like my great-grandfather to pump most of the groundwater to the surface. Moisture sucked from Ogallala sands fueled a feverish transformation of the Plains. The lines of industrial agribusiness covered the land and drained its secrets. Today, the Ogallala aquifer supports around one-sixth of the world's annual grain produce. Withdrawals from the aquifer depths account for one-third of all irrigation in the United States. Almost all of the groundwater is used for irrigating industrially raised crops. The aquifer, however, has long been unable to keep up with the demands, as most areas recharge far more slowly than the water is pumped out. Overuse has taxed the larger High Plains aquifer system beyond repair. Now the timeless groundwaters are running dry. This book addresses the depletion of the aquifer, in all of its complexity and incoherence.

The future of Plains society hangs in the balance. The stakes run deeper than that. Depletion condenses the most urgent conundrums of our times into a single drama. On the High Plains, it blurs the boundaries between the planetary and the personal. In my case, it requires confronting my complicity in the present. The following account is my attempt to reckon with what I find intolerable about the world at a moment when it feels as though something vital is running out. It recounts a search for sustenance among destruction's seemingly intractable roots. The search began in 2016, with a two-year journey back to my ancestral homelands in southwest Kansas.

These returns brought me face to face with the personal registers of depletion as well as its systemic determinants. But I soon learned depletion is not so easy to grasp. It never appeared where or how I expected. Depletion is more than the sum of its apparent parts. It is most elusive in the places it should be most obvious. And it is most clearly articulated when it remains implicit. Its irrationality germinates beneath the surface of seemingly unrelated events. As such, it is most accurately perceived sidelong, from the corner of the mind's eye. It begins and ends as an intimate language passed down over generations.

Among other things, this means many scholarly analyses miss the point. In trying to make sense of depletion, most explanations focus on only one of its many contradictory facets: management, science, technology, belief, historical antecedents, and so on. In doing so, they are unable to convey the beating heart of the issue, let alone alter its rhythms. Depletion always exceeds any single narrative of it. This creates a paradox. Many academic attempts to model, chart, or systematize its causes and effects lend momentum to the expenditure they aim to protest and then end up depleted, too. Best-selling literary narratives that reduce rural personhood to flattened stereotypes and the predictable plot lines of characters known in advance are even worse. How, then, can we account for the overwhelming crises of the contemporary?

Depletion requires its own genre to approximate. This book aims to contribute to such a genre. The structure and content reflect this aim. The narrative approaches its object of analysis by conjuring its form. Like an aquifer, the account is composed of many sediments. Each granular piece evokes a distinct way that depletion is inhabited and the minor potentials for recharge it may hold. The sediments are vertically stacked in layers. They are patchy and unevenly spread. Repetitive themes run between them: memory and amnesia, homelands and exile, holding on and letting go. At times, the layers flow together and connect. At others, they are interrupted and blocked. This pattern of blockage and flow constitutes the book's conceptual argument.

I leave theoretical exposition for other venues. My concern here is to write as closely as possible to the core of an urgent and inchoate problem. That means trying to gesture to the range of apparently disconnected elements that drive aquifer depletion and tie it to other domains of loss and resilience. These elements are at such odds with one another that they wildly exceed any singular analytic register or voice. Pinning down one dynamic allows others to proliferate or expand. It is this contradictory nature that defines depletion as a lived experience and makes it such a perplexing challenge to understand. In what follows, I attempt to invoke how these oppositional elements are held in generative tension in order to subvert narrow accounts of environmental loss and gesture to more sustainable ways ahead.

This allows the book's central question to surface: How can we take responsibility for the future we are now making? It has taken me a long time to realize how prior choices are calcified in depletive substrates that, in turn, exert real force on people's lives. These choices condition our possibilities for finding common ground amid partisan divides, interpretive failures, and eroding democracies. Decisions now being made around the aquifer will delimit future lives in turn, whether we admit it or not. Taking responsibility for what we will leave behind is the book's motive, challenge, burden, and central pivot.

It is obvious that there are no quick fixes. Readers hoping for an easy solution or clear through line will be disappointed. The malaise runs deep. Any pursuit of the aquifer crisis means jettisoning much taken-for-granted common sense about people and politics. It requires drilling down through residues of destruction into the mythic masks of the American frontier and back up to the surface where the waters evaporate into mirages that hover at horizon's edge. The following account commits to these mirages not to dispel them but to more viscerally convey their illusory realities. That means it is equal parts analysis of complicity, elegy for a dry future, unreliable witness, handful of rubble, charter of resilience, and portrait of a lost homeland pieced together from dreams. Through its incomplete imagery I reach out to the ghosts of my ancestors and to the generations to come.

RUNNING OUT

LINES

On the high plains of western Kansas, there is no clear line between water and second chances. Although I didn't know it at the time, I was in search of both when I turned my Prius off a two-lane highway and onto the washboard gravel that led back to the farm.

After fifteen years, the land matched my memories of it. I recalled precisely the vault of space, the circled sky the most dominant feature and the sun a physical weight. Grids of stubble that rotate every half-mile, from corn to wheat to sorghum to corn. Each field a parable about boys who become men by learning to plow every inch, by knowing what not to know, by never leaving or by never coming back.

The road dead-ends at the breaks of the now dry Cimarron River, where the tablelands fall abruptly to a ribbon of shortgrass nestled in a river bend with sage sand hills rising to the south. Here stands the Little Rock House. Named after century-old concrete walls and corrals, it was once my great-grandfather's cattle camp. It is where I spent most of my adolescent summers and it is where my father has returned to live out his years amid broken flints and buried bison bones.

He answered the metal door in his preferred retirement getup, a SpongeBob bathrobe and a silverbelly Stetson. At that moment, I realized the only thing I could say for certain was that he was born to this gusty land. Before, he liked to keep his distance. By then, he had little choice. And sometimes distance is as close to caring for each other as a father and a son can get.

I stayed one night. Like always, I slept in the fixed-up barn, not the house. When the irrigation motor woke me up, I knew I'd been gone too long. The unmuffled Case engine sits a mile-and-a-cornfield from the Little Rock House. The big motor pumps 1,400 gallons of crystalline groundwater per minute nearly every day from early spring to late fall straight into the desiccating wind. Its steady drone was the backdrop to my childhood summers on the farm, as ordinary as the heat and flies and storms. But that first night back, its ceaseless rhythm was distracting.

Because it draws from the water sands under the former riverbed, this motor was one of the few wells in this corner of the Plains still pumping at full capacity. After eight decades of intensive irrigation, other wells in the county had dwindled. Many had gone dry. In 2014, the well at the Little Rock House hit bottom. My father redrilled and luckily hit water farther down. Since then he'd grown increasingly alarmed about the dropping water table. He asked around. Neighbors confirmed his suspicions. During my visit, my father told me about a nearly incredible scale of aquifer decline.

His stories lingered after I left. Over the following weeks, I checked out my father's claims. The situation was worse than he suspected. I learned that southwest Kansas is a front line of the global water crisis.[1] The planet's supply of freshwater is unable to meet the demands that intensive agriculture places on it. It is predicted to only get worse. Pollution and population growth combine to make access to safe water an urgent concern for people across the world.

Groundwater, in particular, is under threat.[2] Worldwide, billions rely on it as their primary source of water. More than half

FIGURE 1

of the water used in agriculture is mined from underground. As industrial extraction grows and the planet warms, these strains on groundwater increase, especially in those dry regions that are becoming hotter and drier. At the same time, groundwater supplies are poorly monitored and managed. Oversight is often nonexistent. Far more groundwater is pumped than can be naturally replenished.

The result is that most of the major aquifers in the world's arid or semiarid zones are rapidly declining. Groundwater extraction is draining aquifers across the globe, including those under the North China Plain, the Arabian peninsula, northern India, central Australia, California, parts of Chile, and many others.[3] Most of this groundwater eventually makes its way to the sea. So much groundwater is pumped to the surface and drained into the oceans that it is now a major contributor to sea level rise, roughly on par with melting glaciers.[4] Aquifers around the world are vanishing. Their disappearance often goes unnoticed or unmourned. Many will never return.

The Ogallala formation is part of the High Plains aquifer system underlying much of the Great Plains, including southwest Kansas. Nearly all of its waters are taken for irrigated agriculture.[5] They are pulled from deep layers of sand or gravel. Owing to this depth, such confined groundwater recharges very slowly or not at all.[6] In most areas of southwest Kansas, this means groundwater is basically a nonrenewable resource. Depletion can permanently alter this kind of aquifer. Once the water is extracted, the spaces between the aquifer sands can collapse.[7] Once the spaces collapse, such sands cannot refill entirely in the future.

Groundwater monitoring began near the Little Rock House in 1958. Since that time, test wells show the water table has dropped more than two hundred feet.[8] Hydrologists estimate that three-fourths of the groundwater in that area is already gone. That is, the Ogallala aquifer has some of the highest rates of groundwater loss in the world.[9] The portions of the aquifer beneath southwest Kansas have some of the highest rates of loss in the Ogallala region. And groundwater losses near the Little Rock House are among the highest in southwest Kansas. All this meant that, by 2016, the place that nurtured five generations of my family was an epicenter of global aquifer depletion.

The paradox is that the loss of southwest Kansas groundwater is coupled with extreme dependence on it. There, aquifer water is the foundation of society. Pumping groundwater has made this semiarid region into one of the world's most productive agricultural zones. The profits of agribusiness depend on subterranean flows. But so do property values, school budgets, job opportunities, and family ties. As experts debate the imminent end of the area's aquifer, state laws allow farmers to continue mining the groundwater that remains. Further depletion is guaranteed.

If groundwater loss is assured, what it means remains an open question. Partisan divides shape how depletion is understood. The loss of groundwater cuts in opposite political directions at once. Some progressives seize decline as proof of rural voters' ruinous shortsightedness. Conservatives invoke it as a rallying cry to protect a threatened farming way of life and to protest regulation. For many Plains residents, groundwater loss is just another pragmatic obstacle that requires grit and hard work to overcome. Regardless of political position, all agree that aquifer loss is a threat to ideal homelands. We do not agree on who can feel at home on the Plains or in the name of what.

The depletion of the High Plains aquifer is a defining drama of our times. Within it, planetary crises of ecologies, democracy, and interpretation are condensed. It demands a response. One return visit with my father made that clear. But I did not know where to begin. Like my father, I was soon struggling to understand the fact that groundwater was running out on the Plains. What else, I wondered, may be running out with it? What would it mean if I could share the concerns of my father and other Plains residents about the declining aquifer? Had we already lost our last best chance at finding common ground? Or was this it?

* * *

On the surface, there appears to be little room for common ground. My great-grandfather RW helped to start deep well irrigation in southwest Kansas in the 1940s. Like others, he imagined it would save the region from Dust Bowl droughts and charter the prosperity to come. For two generations, it more or less worked.

FIGURE 2

Around eleven thousand irrigation wells have transformed this corner of the former Great American Desert into the so-called breadbasket of the world.[10] Agriculture in southwest Kansas generates several billion dollars of revenue each year. Nearly all of the shortgrass has been plowed into fields. Industrial agribusiness has remade this space in its own image.

Grain fields, dairies, hog barns, and feedlots blanket the vast landscape. Grain elevators loom for miles. Dirt is plowed up to front doors. Commodity prices are incanted on the radio. Spray planes pass in formation. Machines never stop. Sprinklers flash all night. Furrows stretch to the horizon. Roads are ruler straight. Mechanical grids and circles dominate the space. When seen from above, the space resembles a pointillist rendition of a bygone land. The drive for order and mastery is etched into every circle and line.

The lines contain their own mythology. Agribusiness myths reduce depletion to economic common sense, as simple as the bottom line. Whether pro- or anti-depletion, abashed or defiant, people often talk about aquifer loss in terms of markets, profit, and arithmetic. The trick is that such logics do not explain depletion but justify and perpetuate it.[11] At the same time, they also deny the realities, contingencies, and capacities of actual people on the Plains. They reduce us all to stereotypes, make eradication seem inevitable, and let everyone off the hook.

The depths, however, may tell a different story. If the surface appears foreclosed by the lines of agribusiness, the subterranean is not. Despite early reports of an "ocean of water" buried under the shortgrass, the High Plains aquifer is not an underground river or lake.[12] It is the remnant of great inland seas, buried millions of years ago by movements of water and wind.[13] Now it is a mass of water-saturated sediments ranging from beach-fine sand to gravels as big as a thumb. These sediments are honeycombed in countless pockets and chambers and bands.[14]

Some of the water-bearing layers are thick and connected and consistent. A well driller's dream. Others are a nightmare. They are thin and broken and fickle. The strata can be laced with clays or stoppered by shales or stretch uninterrupted for miles or stacked all at once. Some of the aquifer's layers breach in sand stream beds. Some are sealed deep and cold. In certain places the layers mingle and communicate. In others, the quiet waters are absent altogether.

The aquifer is patchy and uneven. It is not a single thing. It is a set of age-old relations between blockage and flow and pore. Most of its underground structures are unmapped. They are poorly understood by farmers, scientists, and policy makers alike. That means the ancient waters move in ways that can seem mysterious to all involved. This patchiness was once the domain of the Water Witch.[15] These practitioners of sympathetic magic traveled the Plains locating hidden groundwater with forked wands of peach or willow wood. Water witchery drew power from the mystery of the depths.

Depletion only intensifies this mystery. So do many attempts to dispel it. They say the bottom of the aquifer is more uneven than the top. Some wells respond to other wells miles away. Others are not affected by those nearby. One farmer runs out while his neighbor pumps for years. The relationship between surface and depth varies wildly from place to place, despite no visible differences in the land.

* * *

The depths are personal as well as geologic. Aquifer waters are suffused with sediments of another kind. Traces of destructive processes linger beneath the Plains. Like groundwater, these subterranean layers accrete over generations and imperceptibly bind people's lives to these lands.[16] Depletion brings them to the surface. When I returned to the farm in 2016, I did not realize how far I had drifted away from home, or what it would take to find my way back.

I left the Plains in my twenties. Dreams of wider horizons led me to anthropology, then to New York and South America. I spent 52 months in the Gran Chaco of Paraguay, studying how the destruction of a forest wilderness altered the lives of a recently contacted Indigenous group. During my travels, I learned many lessons and found other families. But I got in too deep and caused damage, too. Looking back, I suppose I was already primed to participate in the violent disintegration of a lifeworld. Something of the destruction I witnessed got on me. I could never quite shake it off.

Anthropologists have developed polite ways to gloss over the insights of return. We talk about reverse culture shock, proclaim ourselves twice born, or repeat T. S. Elliot's adage that the end of our exploring will be to arrive where we started and know the place for the first time. For me, it was not so nice. Call it shame, witchcraft, PTSD, my toe on the trigger—it doesn't matter. I came back out of synch. The overwhelming feeling was something like homelessness.

Shortly afterwards, I returned to the Little Rock House only to perceive traces of the same disintegration everywhere I looked. The Plains appeared to be a mirror image of the tropics. Kansas and Paraguay shared a set of visual and moral cues. The same agribusiness logics were evident in each. It was no coincidence.[17] Some of those deforesting the Gran Chaco were the same people driving groundwater depletion in western Kansas. A group of farmers traveled between the Kansas Plains and the Chaco. Bibles, tractors, seeds, cattle genetics, and market trends went with them. Newly cleared fields in Paraguay, I wrote earlier, looked like pieces of Kansas laid somewhere they didn't belong.[18]

Aquifer loss made this kinship more than a footnote.[19] While my critiques of those responsible for deforestation were unsparing, accounting for aquifer depletion cut closer to the bone. It meant confronting my family's complicity in those very same patterns of destruction. It meant trying to own up to what I was responsible for, too.

In 2016, complicity was on my mind. I was teaching anthropology in Oklahoma, where I had a front-row seat for the social upheavals of that year. Some nights, earthquakes caused by fracking shook the ground. Sometimes on my way to class, I stepped over leaflets from White supremacist hate groups.

Popular elegies were no answer at all. Environmental crisis and the hard-right turn in the United States added a sense of urgency to my growing question. How could we take responsibility for the future we are making?[20] It felt like time to do so was running out.

By then, running out had become an intimate habit of my own. I was in my late thirties, a string of broken relationships behind and an uncertain career ahead. I tried to put down roots and build lasting ties and write a world where I belonged. My attempts failed. Usually it was my fault.

Somewhere along the way, it grew easier for me to cut ties than to repair them. Not long after my return from Paraguay, I broke with my father over an incident that cut me deeply but seemed so minor to him that I'm not sure he noticed. Years passed. I did not know if I was still capable of reconciling with him or with the place my life began. I did not know if I could hold myself to account, either. Searching for the home I'd lost seemed like the only way to find out. I needed to try before it was too late.

These currents eventually came together in aquifer waters and the depletion of them. At the beginning of my search for water, I didn't have ways to truly understand how people come to live with imminent eradication. I was missing the vocabulary to talk about what it means and how it feels when something crucial is running out. I soon learned that depletion is a personal and emotional language, saturated with familiar words.[21] Those words tell stories that remain unfinished. And those stories may lead to other ends. A few of those stories are mine. Like aquifer waters, they seep through layers impossible to chart from surface appearances alone.

* * *

Groundwater Management District Southwest, or Southwest GMD, is spread over the twelve counties of southwest Kansas. It covers around 8,400 square miles of land and 4,400 farms, including the Little Rock House.[22] This area accounts for around a third of the state's agricultural economy and nearly half of the state's groundwater use.[23] If there is an institutional center to aquifer depletion, Southwest GMD is it.

Kansas established five GMDs in the mid 1970s, in response to local concerns about the dropping water table.[24] The original purpose of the GMDs was to conserve groundwater, stabilize agriculture, and allow western Kansas water users to determine their own future destiny.[25] The state ceded much, but not all, aquifer governance to the GMDs.

Today, the GMD boards have extensive authority. They can sell water rights, monitor use, permit or restrict pumping, and grant waivers to exceed legal allotments.[26] According to Southwest GMD's rules, membership is restricted to those who own forty acres of land or water rights to one acre-foot, equivalent to around 325,000 gallons. Only members can vote on its policies. Because my father owns land, he is a voting member and officials may have felt obliged to meet with him.

It seemed like the logical place to start asking questions about decline. When I started learning more about groundwater depletion in 2016, I did not intend to write about my family or the Plains waters. I told my father I was just curious and wanted to learn more.

At the time, I took it for granted that I was on my own. I thought I no longer required anything from my father. I would have said that the disappointments of my childhood were in the past, that I had no expectations left at all. I thought I could make the right choices about what was entrusted to me or suffer the consequences. Like the cowboys of the paperback westerns I read as a child, I thought that I did not need help.

My father tried anyway. He offered to call and make an appointment with someone at the GMD. When I asked him if he wanted to come along, I presumed he would say no. He chose to join. I don't think either of us knew then how far the partnership would eventually go. Only now do I realize his presence was the reason most people agreed to talk to me at all.

We made the hour-long drive from the farm. Over the radio came predictions of a drought for southwest Kansas. We listened quietly until it was over.

Are you ready for this? I asked.

Sure, he said.

You probably know this already. I just want to hear his point of view.

Alright. I got some questions too.

Let me handle it.

We'll see what happens.

My father didn't ask me what I was really after or what I hoped to find. At the time, I could not have given an honest answer. To say I was interested in groundwater or family history was to not say anything. To say I was making a journey to rediscover my home seemed sappy and abstract, even though in some ways it was true. How could I have articulated that I wished to come closer to the mystery of the aquifer in search of some unknown kinship that might be running out with it?

The headquarters of Southwest GMD were in an unmarked brick building off a residential street in a town fifty miles north of the Little Rock House. An official who I'll call John greeted us warmly at the door and ushered us into a boardroom with a shiny table. He agreed to let me record our conversation. Large portraits of the fifteen sitting board members lined the walls. Based on the pictures, they were all ruddy White men like us.

Are these all irrigation farmers? my father asked as we sat down.

Most of them, said John. He wore loafers instead of boots and wool instead of denim. He used statutory metaphors more assuredly than farm ones.

You know, it's sort of a third barn door being open kind of thing, he said. It all depends on how you define conserve.

Let me ask you about that, my father said. I just bought a piece of land without water rights. The guy who sold it to me said the state owns the water and it's going to reach a point where they won't let people irrigate anymore because the water is running out. Is that true?

Well, John said, it's true that the resource has been dedicated by the state legislature to the people, so the people of the state can use the resource subject to a process of application and the putting of water to use for the public interest. So how do we define public interest? It is commerce.

John then walked us through the basics of aquifer policy. I was surprised to learn that most seemed to encourage water use. Before 2004, he told us, Southwest GMD's standard for permitting new wells was based on how much of the aquifer would be lost over a twenty-five-year period.[27] If a well was predicted to deplete 40 percent or less of the groundwater, then it was allowed.

The standard coincided with a state use-it-or-lose-it policy. Until 2012, farmers were required to use their full allotments of groundwater, or two acre-feet of water per acre per year—roughly 651,000 gallons per acre.[28] A farmer who did not use this amount risked losing water rights for the future. At the same time, the state kept granting more farmers the right to pump water. The area of Southwest GMD was closed to further allotments only in 2015, the year before our meeting.

The result was what John called an over-appropriated aquifer. The term was unfamiliar to me. Later I looked it up. It means that farmers are permitted to pump more than the amount of groundwater that remains. Depletion is authorized by law.

Staggering amounts of water are involved. In Southwest GMD alone, farmers are permitted to pump 3.6 million acre-feet of aquifer water per year.[29] By way of comparison, the city of New York consumed less than one-third of this amount for domestic and industrial use in 2018.

This water is taken from an aquifer with negligible rates of recharge. Aquifers are replenished by water moving down through a porous surface or across the saturated layers by lateral flows. Across southwest Kansas, these movements are irregular. Only a few shallow areas of the region's aquifer system recharge quickly. Most refill very slowly. Some zones contain fossil waters that never replenish at all. Aquifer recharge rates near the Little Rock House range from 0.1 to 1.0 inches per year.[30]

That day, we learned farmers in Southwest GMD do not extract all of the groundwater they are permitted to pump by law. Some claim conservation is the cause. But John said the real reason is that the water no longer exists. There is a name for the gap between the actual amount of groundwater and what is legally authorized to be extracted. They call it paper water.

Let me ask you about that, my father said. I'm down by the river, totally encircled by irrigation motors. All of my neighbors double-crop. The water wells never stop pumping all year long. I can't see how that is not overusing their water rights and taking mine. I don't have any way of checking.

We could investigate that, John said. They're not supposed to overpump. But as long as they follow the rules, he said, they

FIGURE 3

have the liberty to implement their strategy to use the water that is allotted to them.

John talked fast and kept going. He told us aquifer decline was necessary for economic growth and he equated the economy with society in general. Because a sustainable level of groundwater use would mean no economic growth, he said, Southwest GMD did not aim to achieve zero loss. Instead, the institution sought what John called a situation of controlled decline.

I was confused. Did that mean the GMD had already conceded the loss of the aquifer or did they have an alternative in mind? John said that an eighteen-billion-dollar aqueduct and water conservation areas, or WCAs, were the best solutions to the problem. These WCAs are effectively gentlemen's agreements between two or more landowners to voluntarily limit their joint water use. Enrollments were very low, although the

WCAs entitle farmers to have greater flexibility in how they use their allotted water.

We have to be careful, John continued, that people don't use the tools of conservation as a way to pump out the water that remains. We need to identify criteria. That's why we're in the process of updating our management process document. Everything is driven by this program management document, he said.

The more I heard, the less I understood. Something did not seem to add up but I was not sure what. On the one hand, it was clear that groundwater management was taken seriously. A large number of experts were actively addressing aquifer loss. On the other, the regional management of the problem did not seem to match its scale. My time in South America had made me suspicious of official numbers and narratives. I felt old habits rising to the surface. It was the kind of mystery guaranteed to catch an anthropologist's attention.

I glanced over at my father's clenched jaw and wondered what he thought. After an hour we were both ready to go. John walked us to the door.

Let me ask you, my father said. What do you see for the future?

John paused. He said he was puzzled by how many people who no longer have any water under their land still don't want to sell their water rights to the government for the common good. For a lot of folks, he said with a chuckle, it seems like giving up known rights to an unknown future is like giving up grandpa.

* * *

Before I started searching for the aquifer, I thought I'd already left my grandparents and the farm far behind. But I learned that destructive inheritances are not so easy to shed.

Groundwater runs through my family lines like blood. My great-grandfather RW was among the second wave of settlers invading ancestral Kiowa, Cheyenne, Plains Apache, and Comanche lands in western Kansas, after genocidal campaigns pushed these Native peoples south to the ever-shrinking confines of Indian Territory. RW followed his own grandfather's travels from central Kansas to the Wichita Mountains and then back. In the 1920s, water drew RW to a sod hut on the flat uplands near the Cimarron breaks, where he dedicated himself to breaking out the windswept range still dotted with skeletons overlooked by bone pickers.

There are many stories about RW. None are about stewardship. Most suggest that he was singularly focused on agribusiness. Someone told me that RW thought he could violate the laws of nature and make money doing it. There is one faded photograph of him holding me when I was a toddler. But I never really knew him. All I could say when I started trying to understand depletion was that he was personally responsible for breaking out thirty-eight square miles of shortgrass, that by mid-life he was the largest landowner in Grant County, and that when he died his eldest son had to sell the farm to a multinational pork corporation to pay his debts. I also knew that RW was one of the first to tap the aquifers below the Cimarron River, where the Little Rock House lies.

There is no denying it. The river is gone and my great-grandfather's wells caused it. Like many desert rivers, the Cimarron was a groundwater stream. Its waters were aquifer waters. The riverbed is a crack in the aquifer's earthen cover. Where the water table was higher than the stream bed, the river and its springs flowed. Otherwise, the waters did not.[31]

For early travelers through the immense grasslands, the only comparable space was the sea. The ground is so level and vast that there is an unbroken view to the horizon dozens of miles away. Traveling over the uniform landscape can give the impression that one has not moved at all, so that the past begins to blur into the present. "That tract may truly be styled the grand prairie ocean," wrote the merchant Josiah Gregg in his 1831 diary entry near the Little Rock House.[32] On the oceanic Plains, waters were mapped like islands. Few of these waters were as precious as the Lower Springs of the Cimarron.

The Springs lay two miles east of the Little Rock House. They were the only permanent source of water in a fifty-mile radius of shortgrass prairie. For millennia, they had been cherished by all those who lived and traveled in the area. Stone spearpoints and butchering sites show that Native peoples inhabited the region for more than ten thousand years. *Cibolero* bison hunters from the New Mexican settlements also frequented the place. In the nineteenth century, the Springs became a noted landmark for those traversing the Santa Fe Trail that led from Missouri to the Rio Grande Valley. Travelers sank a wooden wagon box into the earth and renamed it Wagonbed Springs.

It was a favored spot. "With its delightful green grass glades and flowing torrent," wrote Gregg, "it had all the aspect of an Elysian vale." "We rejoiced," he noted, "at having again made a port."[33] In this place, the Cimarron was said to be a clear stream that flowed year-round through tallgrass meadows and a small green valley. Early visitors agreed that it appeared like "an oasis in the desert."[34]

A nineteenth-century government surveyor described this place as "a series of water holes connected by a narrow channel of pure sparkling water," which he could easily step across.[35] The water trickled into green pools lined with rushes. Bison, elk, antelope, wolves, and bears once drank from the waters. Each year migratory water birds filled the valley. In the pools

FIGURE 4

lived otters and beaver and quicksilver fish found nowhere else. The first settlers recalled waters so rich that "the cowboys would tie their ropes on two ends of a seine and take up enough fish at one haul to feed everyone around."[36]

The Cimarron was notorious for its quicksand and floods.[37] Its fragility must have been less obvious. Records show that plowing of the uplands along the river began in the 1880s but grew exponentially between 1903 and 1914. Eroded topsoil soon choked the Cimarron's flow. In 1914, a massive flood filled the riverbed and valley with silt. Some parts of the river's channel expanded from thirty to eight hundred feet wide overnight.[38] Other floods followed in the 1940s. The clear ribbon turned into a damp muddy sheet.

RW stayed through the Dust Bowl and acquired as much land around the Cimarron as he could. This included the Little Rock House. It was built in 1906 out of crude concrete made with river rocks, a line camp for one of the early ranches. Deeds show the land was sold a dozen times by settlers and speculators before 1945, when RW bought it cheap from the Federal Land Bank. Shortly after, he began irrigating along the Cimarron breaks.

One decade of deep-well pumping was enough to drop the water table some 30 feet below the surface. The state started tracking groundwater levels near the Little Rock House in 1958. By 1976, the Kansas Geological Survey reported that the water table in this spot had dropped by 134 feet.[39] It was the highest measured decline in western Kansas and among the highest in the country.

Irrigation motors meant the end of living water in the river and springs. As the aquifer waters sank, the channels turned to sand. The pools dried up. So did the region's smaller creeks and streams and playa lakes. With them went all the life the desert waters had nourished. The river stopped flowing in the 1950s. The Cimarron vanished around the time my father was born, and so did Wagonbed Springs.[40]

The name of the river still lingers alongside those of other dry watercourses. Cimarron is an archaic Spanish word that refers to a fugitive or something gone feral. It hints at forgotten stories of flight, defiance, and refuge. The river's former path through western Kansas is bordered by sage sand hills that defy the plow. On satellite maps, the sandy bed is a twisting thread from New Mexico's grasslands that disrupts the linear grids and crop circles. Today it contains the last corridors of unplowed shortgrass in the county. The river valley is home to remnant bands of mule deer and prairie dogs and eagles and coyotes.

They say RW was fond of hunting coyotes along the drained riverbed. He would shoot them on sight and poison them and drown them and run them with relays of greyhounds until they died. Those who knew him remember the time he wrecked a brand-new Lincoln car chasing a coyote across the river breaks at night. I picture his fuzzy figure in the bucking car and wonder whether my great-grandfather ever imagined that he would alter the flows and rhythms of this land or that doing so would be the legacy left his descendants. I wonder if he felt any kind of loss as the river dried up and the furrows lengthened. I doubt it.

* * *

If RW offers one kind of settler inheritance, Lila Fern—his daughter and my grandmother—may offer another. She was raised in a clapboard shack on what they called the Old Neese Place, amid prayers and dust storms. Her high-school graduation picture shows a lovely black-haired young woman with an easy smile. By all accounts, Fern was as strong willed as her father. They clashed frequently. She left home and was married in Guymon, Oklahoma, the day she turned eighteen. Her husband Roy came back broken from the war. Against her wishes they returned to the Little Rock House.

Fern spent the rest of her life stuck in a small prairie world under the thumbs of less-talented men. In her mid-thirties, she was afflicted by what they called nervous breakdowns. Afterwards, she grew obsessed with traces of the past. She spent the rest of her life trying to reassemble them.

When I was a child, she walked through pastures to show me Wagonbed Springs and the ruts of the Santa Fe Trail and the grass circles she called tipi rings. I remember her talking about the kind of man she hoped I'd become but I don't remember what she said. She died when I was almost twelve and I have forgotten her last words to me.

She spoke about the Springs so often that they assumed a mythic quality in my mind, although I retain no details of what she must have told me. Fern knew a great deal about the area. She led the county historical society. She founded the town's museum, organized reenactments of historical events and compiled records of all kinds. For several months in 1972, she hosted an AM radio program that shared episodes of local history. But most of her efforts focused on memorializing the original location of the Springs. Perhaps that is why my memory of her is fused with her pursuit of the vanished Springs and why my search for water so quickly blurs into a search for kinship and home.

In middle age, Fern began assembling files of notes for books she never wrote. She dreamed about escape routes and she sought solace in writing a different world. When I stayed at her house I often fell asleep to the smell of her midnight cigarettes and the clatter of her typewriter but I never read anything she typed. It was somehow fitting that most of her files ended up back at the Little Rock House, packed in three battered tin cabinets in the barn.

There I found them in the spring of 2017, almost three decades after her death. The yellowed folders were chewed by mice and sutured with cobwebs. I leafed through them slowly. There were dozens of files labeled in her slanting script with themes like "Crops" and "Churches" and "Indian Stories." Each folder was stuffed with newspaper clippings and correspondence and carbonic transcripts.

In the middle of the second bin lay her thickest folder. "Irrigation" was written on the label. I stopped. Then I eased it from the cabinet and held it in my hands. Time seemed compressed into the fragile tissues and loops of faded ink, hopes and disappointments cycling together like seasons. It felt like some kind of sign.

FIGURE 5

I opened the manila paper to discover a mass of details about pumps and flows and acres, as well as transcribed interviews Fern had done with farmers. On the first page were notes she had typed to herself. "Let's have a little discipline, doll . . . all this stuff in your files and in your head is good for absolutely nothing if you don't get it down in black and white on paper for posterity. . . . You know all this stuff about irrigation: Write it down!"

Fern collected far more than she ever wrote. Her archive traced one genealogy of groundwater. It included the local origin story. According to this story, the area's aquifer was discovered in 1893 when a well driller from Lancashire, England, named William Humphreys struck 240 feet of sheet water while he was drilling a water well for the Johnson townsite, using horses and a wooden drilling tower. Humphreys set off a furor when he declared that he had found a massive subterranean river, with "enough water to float the largest ocean vessel," buried under the nearby Plains.[41]

Others tried to pump this mysterious sheet water to the surface.[42] Fern's files showed that area newspapers tracked early drilling experiments like melodramas.[43] Editors detailed how a farmer's choice of bore width, motor power, and pipe casing were rewarded or punished by amount of flow. The principal characters in these moral plays were all men like Charles Rea, H. H. Brown, Buell Scott, R. R. Wilson, and Earl Brookover. The role of women in farming was hardly mentioned in public.[44] Fern noted the first successful well in Grant County was drilled in 1938 on Otto Fischer's place, the same German settler who poured the concrete for the Little Rock House thirty years earlier. The discovery of a huge natural gas field in the area boosted irrigation in the early 1940s. By 1966, Fern wrote that flood irrigation from the aquifer watered nearly half of the county's land.

When I found her files, I did not yet know that Fern's notes contained few of her own reflections about the groundwater declines that occurred through her lifetime. Those she did give were cryptic: "watertable drop—increased cost of natural gas—question: 39 ft per year, average?" I can only guess how she may have imagined the spread of irrigation and its effects. I cannot say if she ever reflected on the ways the dropping water table mirrored her own personal struggles. But she was undoubtedly aware that her father's irrigation had caused Wagonbed Springs to go dry. I began to wonder what those lost waters meant to her and why she held so tightly to their memory.

The turnoff to Wagonbed Springs lies two miles down a dirt road. It is easy to miss the dusty sign, which points down a double-track road into a pasture. On a blanched summer day in 2017, I unhooked the wire gate and bounced across the cattle

FIGURE 6

guard. I followed the bumpy path for a half mile, past the ruins of an old dugout and stick corral and dead walnut trees to a small clearing of buffalo grass and bitter gourds. A red pipe fence enclosed a square of imported tallgrass. Just beyond it was a circle of irrigated wheat. As I wandered through the enclosure, I could hear the motors on the breeze.

There was nothing about the place to suggest its past. The site looked like any other spot of pasture, except for a handful of plaques and monuments scattered seemingly at random. The words on them were faded by the sun. I supposed one of these markers had been placed by Fern but I didn't know which. A replica wagon box, bent and half full of faucet water, was set into the ground in the middle of the enclosure. Just beyond the pipe fence stood a pink granite marker. It was the same size and style as a tombstone on a grave. The stone angled toward a sheet-metal silhouette of a buffalo bull, set on the bank of the former river. Something about the likeness was off. But dozens of bullet holes pierced the rusted flanks and head.

Brown sand hills framed the strange figure, with the empty riverbed just beyond. There was no stream, no waterhole, no springs, no rushes, no meadow, no fish, no birds, no visitors. A line of dead cottonwoods tracked the dry channel. They had perished when the water table dropped below the reach of their roots. The massive trunks were peeled and bleached a pale gray. Broken branches reached for the sun. They lay like the bones of some extinct creature that once filled this space and enlivened the yellow grass and whose absence was to blame for the austere flatness of the scene.

* * *

Bones tell secrets and I only know a few. Once a librarian gave me a horse. Mrs. Smith worked in the town's elementary school library for a long time. So long that she had blended into the murmuring voices of children and the smells of cafeteria sweet rolls.

I started school there as a fourth grader, after my father sent us away to the Plains for the second time. My sister and I moved with our mother into a prefab home that faced into the north winds. Mom worked long hours to make ends meet. I spent a lot of time in the tiny school library, with its secondhand books and shrunken tables.

They said the only friend a cowboy needed was his horse. When I finished all the horse books Mrs. Smith had on hand, I read them again. Fern must have heard me talking about it. She started to write me cowboy stories and she found people who had horses I could ride. She never told me that she had set it up or whether she paid them. She just took me to some barn or pen where somebody was waiting and they pretended like I would be doing them a favor to work their horse. I never wanted to leave the animals no matter their disposition. Even their sweat smelled good to me.

A year later, my father came back. He had lost my sister's dog but had found a girlfriend with a southern drawl. He moved the woman into the middle of a town where everybody knew everybody's business and got her pregnant and my mother filed the papers. One night around that time Mrs. Smith's husband called the house and asked if he could give me a horse. Now it occurs to me that my grandmother must have arranged it. But I cannot say for sure.

The horse was named Stone. He was flea-bitten gray and grotesquely swaybacked and his brown eyes glowed and when Mr. Smith backed him out of a rusted trailer in the old feedlot where my dad was staying I thought he filled the space like lightning. Just before leaving, Mr. Smith hesitated. He stopped his two-tone Ford and got out and walked over to where the horse and I were staring at each other. The big man bent over and looked at me under his John Deere cap. I could smell his aftershave on the wind.

He said there was one condition for the gift and it was that I do my best to treat the old horse right and that I bury him when he died. The horse had done good for him. He wanted to do good by the horse. I nodded my oversized cowboy hat. I said I promised and took his liver-spotted hand and shook it. He started to say something else but he didn't and that was the last time I saw him.

My father soon moved from the old feedlot to the Little Rock House. My sister and I moved with our mother to a meat-packing town fifty miles south, where the profits of industrial agriculture were built from the labor of Hispanic, Black, and resettled refugee workers. We came back to spend our adolescent summers on the farm. Things happened. Some were ugly.

There are many stories that I could tell about the care of my mother and my sister. Out of respect for their privacy, I'll just say that getting through those tough times sealed the pact between us forever.

During those years, my father worked the night shift at the gas plant. Often our new stepmother wouldn't let us stay in the house with her daughter and the baby. We slept in a camper and found company elsewhere. My sister wasn't too interested in horses but Stone carried me far away.

I rode him aimlessly: up the riverbed, around the Big Circle, down the dirt roads. Eventually the horse wouldn't let anyone else ride him but he did what I asked without saddle or bridle and he grazed quietly while I napped on his bare back, cowboy hat covering my face. I try to recall what I dreamed during those horseback naps but I cannot. Stone outlived Fern by a decade, far longer than anyone expected.

I was in college when the horse died. No one told me. When I heard six months later, I made the long drive to the farm to bury the bones. They said he was in the south pasture but I didn't find the skeleton after two hours of walking and when I turned back to the house I knew it had gone wrong. My father's story started to slip so I dropped it. Later, a laughing stepmother said they'd dragged the body to the road with a chain but pieces kept falling off and the rendering man had to be convinced to take what was left.

* * *

A man I'll call Robert now lives in my great-grandfather's home north of the Little Rock House and farms part of his former land. In many ways, he is a typical irrigation farmer. Our ties are both tenuous and deep. When Robert was a child, RW took him to church and helped teach him to farm.

In the late 1990s, my grandmother's brother Gene sold the farm to a corporate hog concern to pay off family debts. The corporation then resold much of the land under the condition that the buyers would waive environmental rules and permit industrial hog barns. It was the beginning of the pig barns in that region. My father eventually bought back the Little Rock House and two quarter sections. Robert stayed to work some of RW's former fields for a few of the new owners.

These fields included the cornfield across from the Little Rock House. As large as four normal fields, it is called the Big Circle. RW installed a half-mile center pivot irrigation system there in the 1960s.[45] The sprinkler was a Valley Model 1060, renowned for its so-called water drive. The innovative machine did not require a motor or electricity to move around its center pivot. Instead, the sprinkler was propelled solely by the force of groundwater channeled through an intricate mechanism of hydraulic pistons, gears, and rods. According to family lore, this was the longest pivot system in the world at the time.

My father often saw Robert's pickup coming and going from the Big Circle to check the sprinkler water. He wanted to help me learn more about depletion. So he called Robert out of the blue and asked him to stop by the next time he was down that way to talk about irrigation.

The following week, Robert accepted the invitation. It was the first time he had visited the house in twenty years. He sat on the edge of the couch, his leg bouncing in work pants.

I haven't seen you in years, he said to me. Since that time you broke the augur off that grain cart. Gene was so mad he turned blue. I don't even remember when that was.

It was around 1995. I was fifteen years old. My great-uncle hired me to drive a tractor and grain cart for summertime wheat harvest. At the time it just seemed like a job. Looking back, I suppose he was giving me the chance to share this family labor before the farm was sold. I took a tractor driving class but had no idea what I was doing. The first day I scissored the grain cart into a pole and sheared off the bolts that held the augur arm in place. We lost half an evening of cutting time while Robert welded it and I just watched.

Now I can relate better, Robert continued, cause I watch my kids do stupid stuff all the time.

My father told him that I was working on a book about the dropping water table and that I wanted to ask him some questions. His leg bounced faster. He said he didn't have much to say beyond what everybody already knew.

But he did have a story. Like all farmers, Robert knew the water was running out. He'd watched the water table near his house drop several feet a year. In some places he could only keep irrigating by six-packing, or combining several low-output wells into a single pivot system. Sometimes he wasn't sure what to believe. Robert said he had to grow crops to survive and that crops don't grow without irrigation. Like many other farmers and bankers in southwest Kansas, he defined profitable crops as corn. Wheat and milo just aren't worth enough, he said. And if they start charging for water or if corn prices fall even further, I won't be able to keep going.

To keep going is a cardinal virtue, moral and economic at once. It implies endurance, strength, and grit. And it justifies doing whatever it takes. Faced with rising debts and dwindling water, many farmers turned to double-cropping. It usually goes like this. Every fall around October, a farmer harvests yellow corn. After harvest, fertilizer and pesticides are sprayed and the same ground is planted to triticale, a winter wheat-rye hybrid.

The farmer irrigates the triticale seed until daytime temperatures fall below freezing. As soon as it thaws, the water is turned on and the land is grazed until late spring, when the triticale is cut and baled for silage, a fermented feed for cattle and hogs. After the spring triticale harvest, the same ground is plowed and again drenched with chemicals. It is then replanted to corn and watered until shortly before corn harvest.

In this cycle, the earth never lies fallow. It is pushed into an artificial hyperproductivity. Large amounts of water are required, as are repeated doses of chemicals. Some return nutrients to the soil. Others kill weeds, fungi, and insects. Most pesticides are proprietary to a specific agribusiness corporation. The company pairs chemicals with seed varieties genetically engineered to withstand their effects. Farmers say that bare ground is a clean field. One with weeds or stubble is trashy.

Robert narrated this in spare terms. I remembered that Plains language is stripped bare. But there is a lot tangled up in its words. Phrases like "running out" can refer to anything that is dwindling. It is commonly applied to the aquifer but is never limited to water. A bank account, an improvident streak of luck, a person's arrogance, a chance to revisit a bad decision or make amends: these can all run out. The term conveys shades of lost time and moral judgment. It is also used to describe those who leave behind family or home without honoring their rightful obligations to others.

As I sat listening within the old concrete walls, it seemed Robert was invoking all of those meanings at once. Only later did it occur to me that perhaps he was trying to communicate what it meant to shoulder a man's proper burdens to someone he knew would not understand. Grit was a shorthand for responsibility. But what did it mean if responsibility was the same as depletion?

I'm not going to conserve, Robert said. I can't. As long as there's water left I'm going to use it. It won't help anyway if I turn off my sprinkler and my neighbor is pumping all he can.

Brad is one of Robert's neighbors. He lives on the uplands a few miles away from the Little Rock House. He greeted us at the door of his brick ranch house. Plowed fields stretched in all directions. He asked me for a card and looked it over with hooded eyes under white hair and a cap.

PhD, huh? I always heard that just meant Piled Higher and Deeper.

We laughed. He welcomed us in and we sat around his tidy kitchen table. He looked at my father and started to talk about my great-grandfather, who he thought was one of the best to ever do it. Brad got his start spraying for the previous generation of big farmers, and in many ways he had become RW's successor. He farmed several thousand acres and usually double-cropped.

A common practice associated with double-cropping is called prewatering. It consists of irrigating a plot of land before any seed is planted. Most western Kansas farmers say they do it.[46] In theory, prewatering is supposed to increase subsoil moisture to aid future germination. In practice, it is not clear that prewatering impacts yields at all.[47] Most of the water is lost to evaporation.[48]

Some farmers say they prewater just in case it doesn't rain. Others say they prewater as another kind of preemptive strategy. Some farmers pump out all the water remaining in their allotments at the end of the year so they can establish a history of maximum baseline use in order to claim higher amounts when—and if— any enforceable limits are placed on water use in the future.

During our visit, Brad mentioned that two months earlier he had gotten a survey from the state about his attitude toward limits on pumping. He wrote back that he would fight any limits at all by filing a lawsuit. With a wink, he told us he didn't really mean it. In fact, he was willing to accept a 25 percent reduction and he knew he would lose any lawsuit. He hoped the threat of it could buy him ten or fifteen more years of water. It was all that he thought was left anyways.

They used to have big wells everywhere, he said. I remember one guy had two big Chrysler motors and a twelve-inch-bore pipe. He hooked them up and it shot water twenty feet out. Nobody thought twice about it. Hell, your grandpa used to run tail water everywhere. They couldn't imagine it could ever run out.

By then, Brad was convinced that aquifer depletion was inevitable. He thought there was no way to save the water. He said he was a businessman and that using water was a necessary part of good business. People who didn't use water were bad businessmen. I'd like there to be water here for the future, he said, but I'm not sure it can make it.

After an hour, Brad mentioned he had some work to do. My father and I stood up to leave. He walked us through the quiet house to the porch. There, he paused. He turned and spoke directly to me. He said he often thought about the aquifer. He said he liked to imagine it as a group of rivers flowing under the earth, following their ancient courses down from the mountains. Sometimes he stretches his hands over the open bore holes of his wells on the Cimarron just to feel the air pushing up from the deep. He said the moving air smells like rain or a mountain creek. At those times he pictured the buried currents flowing in the dark and if it wasn't recharging then at least he thought that maybe the riverine movement would never stop. I'm not sure it's true, he said. But it is what I like to think.

* * *

FIGURE 7

The evening after our talk with Brad, my father and I walked up the hillside pasture north of the Little Rock House. The old Valley sprinkler lay there like a shipwreck. Chemical spray drift from the adjacent field had killed the shortgrass underneath it. The carpet of buffalo grass was shriveled and brown. On the dead patch sat twelve disaggregated segments of galvanized pipe suspended by wire cables to A-frame towers. The pieces of the water drive rested on the crossbars, the metal stained by aquifer minerals.

I ran my hands along the rusted bars and bleached piston and water hammer and indecipherable gearbox. My father knew the mechanism in detail. He told me how the water once animated all of its parts and turned the iron wheels and how the span crept over the field, and then he told me how much water it required to function. The amount was shocking, in its impossible abundance and that it had once been taken for granted. It could only fit within present reality like a fantasy or a dream. Looking at the obsolete machinery, I thought about the imagined world condensed and mocked in its design. The towers lay still on the dead grass, reaching toward clouds.

We turned back to the house. My father was slowed by injuries from his sports and rodeo days. We stopped at the barn. He made his daily rounds as I watched. He checked water for his cattle and fed the dogs and the barn cats and the goats and a group of three horses. All the horses were too old for work. One was a sorrel gelding that someone had given to my father because otherwise they were just going to let him starve in the pasture. The horse's teeth were so bad he couldn't swallow any of the grass he constantly chewed. The horse had to eat a special soft mash. This meant that every day my father had to prepare the mash and separate the horse and feed him in a pen and let him out and then take the bucket back. I soon learned that my father occupied most of his time with such caretaking. He went through the chores in a fixed pattern. As if in penance. I wanted to help but didn't know what to do or whether he'd care if I tried.

Instead I wandered to the tanks. The horses drank in great slurping draughts with apparent satisfaction and eyed me. I thought of Stone. My father called out to me over the wind.

Hey, do you smell that?

What?

Do you smell that?

No, what?

It smells like magic marker.

He inhaled deeply. Pointed.

Northeast.

My nose was running from the raw wind and I didn't smell anything. I closed my eyes and faced into the breeze and breathed in as much as I could. One breath. Two. On the third I got it, just a whiff of chemical tang. Like a permanent marker. Then it was all I could smell. I opened my eyes. My father was frowning.

What is it? I called out.

I don't know.

There's nothing over there for a long ways besides that gas well and those corrals, right?

Let's go have a look. Maybe it's a meth lab or something.

We got in the truck and took a rifle and drove up the road and then turned down the fence line, bouncing over the rutted trail. It was dark when we turned into the corral lot. We circled it slowly with yellow headlights. There was nothing there but the acrid smell of manure and hay. We turned around. By the time we reached the house, the wind had shifted and the smell vanished.

The dogs barked all night long. I could hear them clearly from the barn. It was funny. Some nights the dogs slept quietly. Most nights they barked sporadically. But sometimes they barked incessantly and strained. On such nights they circled the halo of the yard lamp together, facing the dark but not venturing into it. I never knew if they barked at the night itself or at their imagination of it or at something brooding under its cover that I could not perceive.

As I lay there in the dark listening to the dogs, I remembered a curious memo that I found in my grandmother's files. On February 22, 1968, Fern had interviewed an elderly settler named Pearl about her childhood. Pearl was raised on one of the first ranches in the area. Its headquarters were situated on the banks of the Cimarron immediately next to the Springs. According to Fern's notes, Pearl remembered the Springs and said she thought they dried up because of the growth of the cottonwood trees and not because of irrigation. She recalled the cowboys cutting two-foot chunks of ice out of the frozen river, which they buried in straw and stored in a dugout. Packed in such a

way, the winter's ice lingered far into summer's heat, memory
and portent at once.

What Pearl called the Indian Caves were on the south bank of
the river where the channel cut into a slight bluff. She knew of no
attempt to explore the caves, according to Fern's memo, but they
lay at the foot of the hill where they found arrowheads, artifacts,
and all those skeletons. She recalled flint and beads and steel and
a beautiful black onyx pendant, which she had lost after keeping
it for many years. In the early days, when all the land was in grass,
Pearl said, the settlers were not interested in these objects. It was
not until the dust storms of the 1930s, she recalled, that people
began to think about gathering Indian bones and relics. She was
unsure whether it was the dust clouds that imbued the bared
remains with their power or if it was the other way around.

Fern noted that Pearl then recounted what she called the old
legend. If you want to know where the gold is buried on the
Cimarron, she told Fern, you go to the Indian Caves and walk
inside and sit there till midnight. An Indian chief and nine braves
appear and tell you the secret. Then they kill the intruder.
Around the time the dust blew, Pearl remembered, a young man
from a nearby town tried it. He was too afraid to go into the caves
but sat in his car nearby with the windows up and the doors
locked. At midnight he heard something and turned around. He
later swore, she said, that in his back seat was an apparition that
he called an old squaw. Whatever he saw, he ran off on foot and
left behind the car and a gold pocket watch that he used to keep
the time. The next day he went back to retrieve both.

I lay there with my eyes closed. I began to picture the vaporous figure of Fern searching for sustenance among the dormant
forces of the ancient aquifer, but before the image was formed
fully in my mind I drifted off to sleep among the barking dogs.

BONES

The next morning we got up early. As my father went through his chores, I thought about the ways irrigation is presumed to mitigate the risks of transforming the Plains into an industrial landscape and the vengeful spirits that depletion may create or unleash.

While some farmers colluded with eradication, others said they preferred to save the groundwater that was left. But they told me they were forced to keep irrigating because of the way farm finance works. To learn more, we went to see a man at the local bank. He ushered us into a boardroom and reverted to high-school hierarchies with my father.

After exchanging pleasantries, the man told us that the bank was not focused on conserving groundwater. Instead, it was interested in what he called pragmatic economics. That is, the aim of the bank was to make money out of agribusiness. Loan counselors do not tell farmers what to do. But they often encourage farmers to follow the lowest risk option and assess their loan applications accordingly. In many cases, this meant the bank urged farmers to continue to grow irrigated corn.

The rub is with the arithmetic of Plains corn farming itself. It is anything but a stable equation. Take 2017. In that year, the statewide average yield for irrigated corn was 197 bushels per acre. To get and process that yield, a farmer had to pay for seed, fertilizer, herbicides, pesticides, natural gas for irrigation motors, equipment repairs, harvest costs, insurance, and taxes. Each farmer guards expenses like trade secrets, so it is hard to pin down exact costs. But according to the Kansas State *Corn Management* guide, the statewide average cost for producing the average yield was $710 per acre. If true, that means a Kansas corn farmer had to earn at least $3.60 per bushel just to break even.[1]

On the arid Plains, this balance is further strained. There, farmers said costs varied wildly and commonly reached $800 per acre. In those cases, they said, less than $4 corn is a losing proposition. Any repeat spraying or major catastrophe or big investment increases the burden. Yet corn prices had fallen by half since the ethanol boom set record highs in 2012 and investors from Iowa bid up western Kansas ground.[2] In 2016, the regional Garden City Coop paid between $3.20 and $3.42 per bushel. The year after it paid between $3.28 and $3.64. This coincided with a severe, months-long drought. Under these conditions, growing corn was like betting against a stacked deck. Farmers were more likely to lose money irrigating a crop with aquifer water than they were to turn a profit.

Debts grow alongside the water-thirsty corn. In 2018, a quarter of all Plains farms reported being indebted.[3] This rate was higher for Kansas farms and higher still for Kansas corn farms. Western Kansas corn farmers shouldered more of this debt proportionally than those in the eastern two-thirds of the state. Weather and markets change the equation. The boom-and-bust story is familiar. It is no surprise that so-called family farms face challenges on the Plains owing to the structural investments and terms of corporate agriculture.[4] What confused me was how losing money on corn seemed to result not in less water use but more.

Later, I asked one farmer about it. He responded as if it were obvious. If you gotta get a crop out before it burns up, you turn on the water and worry about it later. That is the only way to get

FIGURE 8

260-, 280-bushel corn. If you can get that, well you're doing alright even with these prices.

This usually leads to overpumping. The practice was reportedly widespread in the recent past, when big wells were common and regulation absent. That began to change in the mid-1990s, when the state required all farmers to report their usage and fit wells with a water meter. Meter tampering quickly became a local industry. Some installed the meters incorrectly, others clogged them up, and a few became specialists in altering the readings. Others were willing to face the punishment for using more water than they were allotted: a modest fine and your name printed in the local paper. In response to requests from farmers who followed the rules, state regulators sharply increased monitoring and penalties in 2017.[5]

If debt compels waste, it also promises a perverse kind of stability. Debts keep the corn moving and keep farmers coming back to the bank. Federal farm insurance subsidizes this debt-waste cycle. It protects everyone from devastating loss. And it can play certain tricks with the linearity of the bottom line.

* * *

My father's buddy Jim is a third-generation farmer who knew about the crop insurance policies sold by one of the government's private contractors. My father called and asked if we could ask him questions about water and insurance. We found him and a hired hand named Wallace shoeing a horse behind his house, surrounded by empty Coors cans at midday. They took a break to talk. Jim walked up, buzzed and handsome in a straw hat with yellow wrap-around glasses over stone-washed jeans. You're starting to look just like your old dad, my father said.

We sat around a concrete slab on plastic chairs. Wallace's wife Amber came out of the barn. She put hot dogs on a grill and Jim poured gas on red ants that denned in the seams of the slab. My arms burned in the sun. My dad asked Jim to tell me about what is known as insurance farming.

Once Jim started talking he didn't stop. He said it was common for folks to plant irrigated crops they know will fail in
order to receive insurance payments. Some farmers put out a
ridiculous crop and don't spend much on it and if it doesn't
work out they still get the payment. The insurance company is
obliged to pay unless it can prove poor farming practices. And
how the hell are you going to prove that? Jim snorted. Most
farming practices out here are poor, right?

The insurance companies, he went on, don't seem to mind
paying with federal dollars. Their real profit comes from high
premiums, which are also offset by federal subsidies. The payment only covers three-fourths of the value of the crop.[6] Jim
said the game was in how this value was calculated. It was not
based on current conditions but was an average of that farmer's
harvests over the last ten years.[7] Those who had run out of
water in that time were reimbursed for yields that had since
become impossible. Under certain conditions, it meant a failed
irrigated crop was worth more than a successful one.

The difference between irrigated and dryland crop insurance
was key. Prior to 2018, when federal officials pioneered "limited
irrigation" insurance to address declining groundwater, these
were the only two options for Kansas crop insurance policies.[8]
Because nonirrigated crops were considered lower value and
higher risk, they were far more expensive to insure.[9] On the
other hand, if farmers pumped water for an irrigated crop they
spent less on the premium and made more money in reim-
bursements if it failed.[10]

And it is big money. In the two counties surrounding the
Little Rock House there are roughly 520 farms.[11] Over the decade between 2008 and 2018, these farms received around $155
million in indemnities to offset lost crop values.[12] Certainly,
most of it was honest. But Jim stated the obvious. Until they

quit insuring corn, he said, people will water it to grow insurance money.

He emphasized that he didn't blame them. He himself farmed six quarters of ground. What I do is just farm the government too, he said. I try to get in every program and get every dollar I can. Why would anybody just quit pumping? It makes no sense.

We sat squinting in the sun and listened to Jim talk. He shifted seamlessly from insurance policies and water issues to snakebites and a land scheme and old pranks he'd pulled with my father. He asked my father if he remembered the bucking machine they'd made and how they could never ride it. It was a homemade contraption with an off-center wheel they dragged behind a truck. Your dad used to laugh at us, replied my father. Boy it just tickled him to watch us get thrown. Jim grinned and grabbed a beer and offered his prediction for the future.

Bleak is the right word, said Jim. I figure it is like those old mining towns you see up in the mountains. They took what they want and when it ran out they left. The water is going to run out and there is nothing we can do about it. There's no utopia out here.

We finished the cold beer and hot dogs. It seemed like Jim wanted to get back to work and my dad was ready to go. I kept my thoughts to myself. We threw away the paper plates and stood up to leave and shook everyone's hand. Amber then asked my father if he ever used to go up to Lake McKinney, about fifty miles north.

A time or two.

We used to go up there to water-ski all the time. Then it got too shallow to ski. Then after that it got so shallow you could go fish by hand. Little while later it was just a puddle and those fish were trapped. I don't like to think about it.

Well.

Just like Bear Creek.

Used to have water in it too.

You should ask Jeffries about that. I bet he remembers it.

My father gave me the truck keys. We pulled onto the road. Man, it is good to see Jim smile. He looks just like his old dad did.

* * *

On the way back we passed mile after mile of cornfields. I tried to reflect on what all this meant, but as usual I was numbed by the details and soon lulled by the lines in the passing crops. From the window of the moving truck I noticed an optical illusion I had seen a couple of times before. At lower speeds the crops blurred together. But once a certain speed was reached the precision and regularity of the pattern allowed thousands of rows to flicker back into focus as the image of a single plowed field floating through space and time. I drove faster on the straight road and the truck started to shake and then my father interrupted my thoughts and told me to slow down or we would get stopped by the sheriff he always sits out here behind that barn. The big white barn sat in the faint depression of the plowed creek bed with cumulus clouds above. As I think back on this image of the barn something about the composition of the scene, maybe the roofline against the slope, reminds me of the barn at the Little Rock House as viewed from the east with the hillside beyond it where once I found a buffalo bone.

I cannot say exactly when it was but I was around eleven years old and the field above was still flood irrigated. Overflow had carved a twisting channel down the middle of the road about a foot wide and a foot deep. The channel looked like a miniature river complete with falls and bends and banks. Something round bulged from one of the tiny bluffs. I nudged it with my boot toe then squatted down and dug around it with a weed stalk. Stained reddish brown, it was about eight inches long and three inches thick with a jagged break on one end. I didn't know what kind of bone it was. Horse? Woolly mammoth? Dinosaur?

I carried the piece back to the house and washed it in the bucket under the faucet and showed it to my father. He didn't know either so he mailed it in a shoebox to the museum at Fort Hays with a note asking them to identify it. Sometime later he received a letter in reply. I distinctly recall the university letterhead and the typed message stating it was the femur of a Plains bison that perished sometime in the mid- to late nineteenth century. The bone was not returned. My childhood thoughts raced on. For years, I had utterly no memory of it at all. It was only two and a half decades later, when I happened to cross the spot on the hillside where the channel had been, that I remembered the piece of bone. Once I recalled it, the bone began to frequent my thoughts. The more I thought about it the more mysterious it appeared.

It was the only trace of a buffalo I encountered. What made it even stranger was how little I knew about the events that erased the bison and dispossessed the area's rightful Native inhabitants and which somehow anesthetized me to their absence. Like all Kansas children, I learned the flattest of outlines about genocide and bison in school. Once there were millions of buffalo, most were killed, a few survived: it was okay. Apparently I had no problem reconciling those imagined scenes with the reality I occupied wherein such beasts were obviously impossible and there was no room for their multitudes save that already reserved for the fantastic creatures that migrate through a child's mind. I did not grow up thinking that something was missing from this story. The bone suggested otherwise.

It is hard to form any adequate idea about the destruction or to grasp the attitudes that made it possible. We can reconstruct, as historians have done, the immense size of the herd, its cleavage into two groups, the gradual fragmentation and confinement of the southern portion, the illegal but sanctioned pursuit of it first to southwest Kansas then to the Texas panhandle, and the final extirpation of its remnants. It is undeniable that the extermination of the bison was an official part of federal wars against Native peoples, that genocidal campaigns and massacres of Indigenous peoples accompanied the eradication of bison and the near extinction of Plains wildlife, and that all of this occurred in a very short period of time. Sources agree that, over a single three-year period between 1871 and 1874, between three and seven million bison were killed within a hundred-mile radius of the Little Rock House.[13] Common accountings of these cataclysmic events seem to require an amnesia that seals them off from the present and is somehow tied to my alienation from those lands.

Looking back, perhaps it was the sense of dislocation conveyed by the bone that compelled me to try to reassemble in my mind the buffalo that perished on the hill from the inside out. I was not able to say why this thought captured my imagination. But I suppose I must have been searching for some ligature that tied me more firmly to the creature, to the place we shared and to the peculiar ways the conditions of its destruction were mirrored within aquifer loss.

I learned that 1872 was the pivotal year for settler colonization in southwest Kansas. The land was surveyed for the Santa Fe Railroad, whose tracks raced west down the Arkansas River valley from the camp called Buffalo City to Granada. There construction stopped and hundreds of track workers were laid off. They found a new market for buffalo hides.[14] Thousands of hunters were involved. Some wrote memoirs. Frank Mayer said he came to southwest Kansas because he felt restive. He heard buffalo were walking gold pieces. "I was young. I could shoot. I needed adventure," he wrote. "Wouldn't you have done the same thing in my place?"[15]

Western Kansas went what Mayer called "buffalo wild." He said this wildness was like a gold or uranium rush. "Men left jobs and families and sold whatever they had to get into the game. I did it myself. And why not? It was a harvest. We were the harvesters."[16] Many like Mayer were also novices. Most harvesters did not know how to shoot or skin efficiently. Lieutenant Colonel Richard Irving Dodge, in charge of army troops stationed near Buffalo City to ensure that no Native people remained north of the Arkansas River, famously estimated that each hide sold represented five animals killed during that first year of madness.[17]

Shooters soon learned. Bison herds were matriarchies. If the lead cows were shot down first, in the order of their rank, the others did not run. They clustered. The habit made them easy targets.[18] Shooters bragged about how many they killed in a single sitting. A man named Tom Nixon claimed to have shot 204 in one stand, part of a thirty-five-day period in which he reportedly killed more than 3,200 buffalo along Rattlesnake Creek.[19] Mayer said his regular quota was twenty-five skins per day.[20] Individual hunters shot tens of thousands over their career. "Adventurous?" Mayer wrote. "No more than shooting a beef critter in a barnyard."[21]

Gun companies designed new weapons to kill bison. The most famous were the so-called Sharps Buffalo rifles. In 1872, Sharps introduced a .50-caliber cartridge. Three inches long, it used 100 grains of black powder to shoot half-inch-wide bullets that weighed 450–700 grains and could kill up to a mile.[22] For his Sharps, Mayer ordered 20× telescopic sights with stadia marks from the Vollmer factory in Jena, Germany. He swabbed his barrel with a mixture of graphite, tallow, and his own urine. He cast his bullets at a ratio of sixteen parts lead to one part tin and handloaded his brass shells with gunpowder shipped from three London companies.[23] Most of the meat was left to rot. One settler who wanted to sell hams fed his hogs with buffalo. A fraction was shipped east as so-called "Texas beef," sometimes tongues were smoked or pickled.[24] The hides themselves were tanned for industrial machine belts and European military accoutrement.[25] Everyone seemed to take the abundance for granted.

By autumn 1873, after just two full seasons, western Kansas bison were almost gone. Dodge rode south into Indian Territory and back. "Where there were myriads of buffalo the year before," as he put it, "there were now myriads of carcasses. The air was foul with sickening stench and the vast plain, which only a short twelvemonth before teemed with animal life," he recalled, "was a dead solitary putrid desert."[26] That year, one railroad engineer said it was possible to walk a hundred miles of the Santa Fe track by stepping from one carcass to another. "The heavy guns rumbled and boomed hour after hour," a Civil War veteran recalled, "just as if a war were being waged."[27]

Hunters set up camps around all the water and harassed the remaining bison. For the last few years of their existence, as hunter John Cook put it, "there were multiplied numbers of big and little outfits camped at most available freshwater places, ready to bombard them wherever and whenever they came in sight. They were kept on the go," he said, "and when they would find a place that was free from the fusillade of lead from the big long-range guns, they would rush and crowd in pell-mell, crowding, jamming and trampling down both the weak and the strong to quench a burning thirst." Many of the buffalo, Cook surmised, were "rendered insane from their unbearable thirst."[28]

The Cimarron was the final refuge of Kansas bison. Comanche, Kiowa, and Cheyenne all knew it as the River of the Buffalo Bull.[29] The Medicine Lodge Treaty of 1867 prohibited White settlement and hunting south of the Arkansas. The area, which includes the Little Rock House, was supposed to be reserved for Native tribes and for the remaining herds. Buffalo hunters violated the law. In 1873, the government moved the so-called Dead Line south to the Buffalo Bull River.[30] That same year, Billy Dixon's group set up a winter hide camp at Wagonbed Springs. "We prospered," Dixon noted, "as buffaloes were plentiful."[31] When hunters asked Colonel Dodge about the penalty for violating this new border, he told them to disregard army policy. "If I were a buffalo hunter," he said, "I would hunt buffalo where buffalo are."[32] "Kill every buffalo you can," Dodge famously stated, for "every buffalo dead is an Indian gone."[33] With this endorsement, the next spring a party of fifty hunters invaded No Man's Land and the Staked Plains. During that summer of 1874, the hunters joined army troops in attacks that aimed to erase all Native presence from the Texas panhandle. Hundreds of Native men, women, and children were killed in what was called the Buffalo War.[34] At the same time, the hunters were slaughtering millions of bison. In 1869, the southern herd was estimated between ten and thirty million animals. By 1879, most Native survivors were confined to reservations and nearly all of the bison had been exterminated.[35]

Straggler buffalo appeared near the Little Rock House through the mid-1880s. They were shot on sight.[36] Fern talked to a man who recalled the final bison in Grant County. "It seems

FIGURE 9

the very last buffalo to be seen here," she wrote, "was a mangey, solitary old bull that a man named Billy McRae captured along the Cimarron in 1890."[37]

Bison were not the only nonhuman casualties. Hunters killed a comparable number of antelope.[38] Pronghorns and elk and deer and lynx and grizzly bears and black bears and mountain lions and beavers and turkeys and ferrets and otters were all extinct in western Kansas by the 1890s. Wolves were gone by the 1930s, if not before.[39] It was common practice to lace bison carcasses with strychnine. After the herds fell, several hundred hunters became full-time wolfers. Mayer called them a "mean, ugly, cheap breed of drunkards."[40] Ravens lived with the buffalo and wolves. One hunter recalled poisoning a carcass in the evening. The next morning, he wrote, it was "surrounded by a field of fallen ravens."[41] The drive to eradicate native wildlife did not stop. Photographs from jackrabbit roundups in the 1950s show men, women, and children coming together in a crowd to encircle and club the jumping animals.

Like irrigation farmers, buffalo hunters used the bottom line to interpret their role in eradication. "But I was no Indian," Mayer wrote. "I was a businessman. And I had to learn a businessman's way of harvesting the buffalo crop."[42] Hunters like Dixon, Cook, Cator, Jones, Brown, and others each wrote the same thing. "Buffalo hunting was a business," noted Mooar, "not a sport."[43] They considered it a rational economic enterprise. What are we to make, then, of the fact that these same hunters eliminated their supply and that they did not profit greatly from the slaughter?

Many lost money killing buffalo. Some broke even. The best earned the equivalent of a shopkeeper's salary. This is according to their own accounts. Dixon, Cook, Mooar, Cator, Jones, and Brown all concurred. Outfits were expensive. Prices dropped. In 1872, a bull hide was worth $3.50 in Dodge City. By 1874, the market was glutted. A bull hide sold for $1.15 and a cow for 65 cents.[44] Dealers in Texas hide camps like Rath City paid even less. Consider Mayer. He started in 1872 and hunted buffalo for nine years from Kansas to Texas to Montana. He said he broke even the first two and made $3,124 profit in the third.

"Then," he wrote, "I slid down to nothingness. I am quite confident that I was among the highest rewarded five men on the range. I have since talked to a dozen of the runners that I knew and all remarked that I got more out of it than any feller they knew. So, I am quite safe in my surmise that a good high average of all runners engaged was less than $1,000 per year net."[45]

Shooters did not make much and skinners made less than shooters. The government collected no tax.[46] The northeast tanning factories destroyed the old growth hemlock forests and went bust too.[47] Mayer asked himself if the slaughter could have been otherwise. "I don't think it could have been," he wrote.

"I think the slaughter of the Buffalo would be what moderns call an historical necessity. It just had to come. But maybe I am just rationalizing. Maybe we were just a greedy lot who wanted to get ours and to hell with posterity, the buffalo and anyone else, just so we kept our scalps on and our money pouches filled. I think maybe that is the way it was."[48]

Newly arrived settlers likened the bone-strewn Plains to a field of white lilies or a charnel house.[49] Bone dealers paid between $4 and $12 per ton at the railroad stations. Poor homesteaders called the bones "a godsend" and "a wise dispensation of Providence."[50] They broke the skeletons with hammers and axes. It took around one hundred skeletons to make a ton.[51] Kansas settlers reportedly earned more than $2.5 million on bones from 1868 to 1881.[52] Immense piles sprouted along railroad tracks. In 1886, the bone pile at Dodge City was said to be a quarter mile long and higher than a boxcar. James Cook said he saw a rick of bones laid by the tracks in Granada that was twelve feet high, twelve feet wide, and a half mile long.[53]

The bones went to carbon works plants in Saint Louis or Chicago, where they were turned to bone ash and bone black and Dippel's oil and used for fertilizer and bone china and pigments and for refining white sugar.[54] A man nicknamed Buffalo Bones shipped three thousand carloads from Dodge City by himself.[55] Other Kansas bone pickers went mad. One named Stewart was roped and tried before a town jury and declared of violent manner and unsound mind. His effects at the time included one lot of bones at Aubrey, one lot of bones at Pierceville, and one lot of hides worth $27.35.[56]

Colonel Dodge noted that buffalo hunting was associated with a syndrome that he called Plains insanity.[57] It began when a person lost all sense of direction in the wide space. The disorientation grew until it overwhelmed the capacity for reason. For

FIGURE 10

someone so afflicted, Dodge wrote, "everything appears changed and unnatural; the most ordinary events appear to possess unusual significance; the nerves become unstrung, and the man soon loses control of himself entirely." A primary symptom was the drive "to keep moving." Dodge said the remedy was to make camp, cover your eyes, and stay in one spot until your bearings returned. But many, he noted, never recovered. They just kept on going.[58]

One of the only photographs included in Fern's files shows a massive rick of buffalo skulls stacked along train tracks in western Kansas. The pile was incinerated a century before I was born. I have no firsthand experience of the slaughter except for the emptiness it left behind. Yet when I see my grandmother's picture of bones or read the hunters' memoirs I have an overwhelming feeling that this destruction still trails me. It belongs to a world that is utterly alien and at the same time intimately familiar. Any attempt to reconcile my imagination of the devastation with images of my adolescence makes my head spin. I cannot say why this tension affects me so profoundly, or what exactly I search for as I turn these scenes over and over in my mind's eye, unless it is some way to come closer to home.

Many of the hunters seemed to be searching for a home within destruction too. Their stories drift like dust. A man named Hoodoo went on one of the last buffalo hunts in the Cimarron country. He roamed the country for weeks and found nothing. His five companions turned back to civilization. He kept on alone. Two days later, he came upon a playa lake in the middle of the Plains. It was like a dream. The seasonal waters were moving with ducks and geese and migratory waterfowl of all kinds. Herds of wild horses grazed nearby. All around the lake he could hardly see the grass for it was covered with buffalo. This scene of the pool and water birds and ponies and buffalo burned into his memory. For the rest of his life he considered it the most beautiful thing he ever witnessed. It was at this lake that he made what he called his best killing. In ten days he killed more than eight hundred of the buffalo. Shooting was so good he fouled his rifle and ran out of ammunition. Hoodoo went up to Granada for supplies and skinners. Others backtracked his wagon and by the time he returned to the lake all the animals were killed.[59]

One summer when I was thirteen, a lone antelope appeared along the Cimarron about a mile south of the Little Rock House. I spent hours crawling through the sage downwind so I could see. The antelope had made it back to these Plains, the ancestral range of her kind. But the journey had cost her. She looked bad. Her coat was patchy, her stomach distended. She did not move or eat or drink that I could tell. The fifth day I crawled to my looking spot she was gone.

* * *

No thoughts of buffalo or antelope were in my mind the day my father and I drove to a public meeting of Southwest GMD. I presumed the future of the aquifer would be debated. The meeting was held fifty miles south of the Little Rock House, in the meat-packing town where my mother moved after the divorce and where my sister and I went to high school. The parking lot was full of white pickups, most with decals on the doors.

The civic center's soaring ceiling and polished concrete gave it the air of a megachurch. We arrived late but it seemed like we hadn't missed much. At the door a young woman handed us each a red raffle ticket. About three dozen people dotted two hundred plastic chairs arrayed towards a podium and screen flanked by twelve board members. Three women were in the room. All the other participants were White men who appeared to be over fifty. It was a jarring contrast to the town itself, where around half of the official population is listed as Hispanic. While the industrial centers of southwest Kansas are racially diverse and markedly multicultural, rural farm management remains a distinctly White, masculine space.

We poured ourselves Styrofoam cups of weak coffee and sat down to listen. The microphone boomed in the cavernous space. An election was underway for two seats on the Southwest GMD board. These were three-year positions. Whoever held them would play a key role in the decisions that were likely to determine the fate of the aquifer. There were only two candidates. Both had served on the board previously. Both ran unopposed. Both won with the unanimous vote of those in attendance.

John, the official, played the part of emcee. But the audio system kept cutting out. In a voice that switched between blaring distortion and barely audible tones, he thanked the voters for their participation and it was his pleasure to introduce the first Technology Minute. Apparently, the annual meeting had corporate sponsors? My father looked at me with one raised eyebrow.

A man in a white polo from Kansas Corn kicked it off and said that his group spent three hundred thousand on water conservation last year. I had never heard of the group. Was it a corn lobby or something else? The man then called out three numbers and raffled off a branded T-shirt. He was replaced by an identically dressed guy from Cropquest, who told us that their water meter smartphone app creates a full season unit of imagery and was now available, and then he called out three numbers and handed out a Cropquest hat.

The Technology Minutes were followed by an expert from the Kansas Geological Survey (KGS). Later, I learned this group was a national leader in the wider struggle to monitor aquifer depletion. The expert said that KGS had 429 test wells across southwest Kansas that could be monitored in real time. The challenge was how to consolidate those measurements into an improved model of the aquifer across the district, one that could generate more precise averages for decline. What we want to know, he concluded, is how the aquifer behaves over time. It is still an open question.

As soon as he finished, John interjected. He was so glad to have the KGS present at the meeting because they were charged with presenting nonbiased information that had nothing to do with policy or politics. The KGS, he said, was a state agency focused only on making Kansas a leader in hydrology. Before I could reflect on the implicit subtext, it was time for another Technology Minute.

A cartoon dragon appeared on the tiny projector and a man said that his irrigation company based right there in southwest Kansas had invented the most precise application of water through a pivot ever. It was being adopted in the Middle East. He raffled off a T-shirt with the logo. Following him was a young man who worked for Servitech, whose Expanded Premium Services included the Profiler Plus, a wireless platform for connecting multiple sensors through a single gateway that you can monitor from the comfort of your pickup or home, and he gave out a Servitech duffel bag.

So it went. I was in over my head. I could barely keep up with the acronyms and institutions and the questions raised by the muddle. Maybe it was my ignorance that made it increasingly difficult for me to tell them apart. I glanced over at my father. His face was hard under his Stetson and he looked straight ahead. I got us some more bad coffee.

A professor of agricultural economics discussed changes to section 1999A of the Farm Bill and besuited personnel from two congressmen's offices spoke about their support for agribusiness and a McCrometer rep told us about new cloud space bubblers and digital flowometers designed especially for low flow situations and the president of another unfamiliar advisory committee almost shouted that the real problem was that the state needed to credit producers who were already conserving water and give them the right to use more and we heard that uranium concentrations in municipal water taken from the Arkansas were twice the Environmental Protection Agency's limits and that we could use Hydroresources's new precision drilling techniques and how the new FieldScout moisture probe with Bluetooth connectivity can upload measurements to your smartphone and how about that gift certificate look at that.

The screen was too small to see. They kept having trouble with the sound system. The speakers' phrases were chopped and remixed. Distracted, I read the program and then scanned the room again. With a jolt, I realized that almost everyone was actually a paid participant. By my count only seven people in the room were not board members, sponsors, or presenters.

When I tried to tune back into the words, I found that I could not. Whether it was due to the distorted sound or the overwhelming rush of details or the implications of the absent public, my ability to register the reality of the scene began to slip. As the meeting dragged on, the strangeness intensified. I felt as if the shuffling of the speakers was perceptibly speeding up, like a deck of cards, until they would begin to fuse together into one faceless, formless machine, and the feeling grew that this machine was united in a singular purpose that could not be spoken aloud but to which those present had already sworn a secret allegiance as if through a compact or an oath. As I try to remember the meeting, oddly enough, it is the memory of this blurriness that stands out with the most clarity in my mind.

Two moments punctured the strange mist that otherwise envelops my memory of the event. One was offered by an official from the Division of Water Resources. In flat tones, he told us his office had increased penalties for overpumping. He stated that there are three legal mechanisms to reduce water use. One was successfully implemented by a neighboring GMD. He said of course we know there is no recharge. Water pumped today is taken from the future. Strict limits, he stated, were obviously necessary. It felt like a relief and a warning at once.

John's concluding address followed and was the other exception to this peculiar haziness. While it is impossible for me to remember most of the details conveyed at the meeting, I am able to reconstruct parts of what John said almost verbatim. Reading my notes of the event, I recollect how he stood at the podium in a brown tweed coat and how his voice sounded as he spoke. He talked fast and smiled often and went ten minutes overtime.

The most pressing issue for water management, he began, was the need to come up with terms that we all understand. Things have gotten confusing for folks. Less water use is not the answer, obviously, because the less water you use the less value your land has. Sure, legally a neighbor can shut another one off but is that in the best interest of our community? Most of the confusion, he assured us, is caused by how we talk about depletion. Talking about impairments caused by depleting water, for example, is confusing. So let's speak clearly, he said. Instead of impairments, he explained, we need to talk about drawdown allowances. These are simple to calculate, he stated reasonably, if we use credits to make a multiyear calculus that can distinguish which wells are critical and by critical he meant those that exceed the parameters of acceptable drawdown, which of course, he explained, is a much better screening tool than impairment so let's try to use drawdown allowance instead of impairment. Another term that would provide clarity and resolve the confusion that seemed to exist among so many of us, he reasoned, was to not talk about aquifer loss in general terms but to speak instead about economical drawdown constraints, which is simply the percentage of saturated thickness that can be lost until any well sinks below the threshold of economic viability. In the absence of better data, we can assume that 70 percent loss of saturated thickness is the same as the loss of economic viability of that well. Once that economic threshold is clear we can then proceed to clarify how we speak about the ways the physical constraints of one well change relative to its effects on neighboring wells, which of course means we need better data on well construction and wall thickness if we hope to project future drawdown vectors with more accuracy while making sure we do not confuse the maximum economic drawdown with the maximum physical drawdown. These are the kind of terms we need to make the distinctions that are required. Such clarity, we all know, could go a long way to prevent lawsuits like the problem over in Haskell County which wasn't good for anybody, of course. It is just as clear that the real long-term answer to our problem is water transport, which could easily be included in the infrastructure bill now in Congress and which would bring us all the water we need to flourish. It is obvious, he declared, that production agriculture in western Kansas is the best in the world. We have everything, the infrastructure, the soil, the labor force, the best people. We only need one thing. Water. The issue is simple, he concluded. There is no magic here. It is just markets and value and partners. Now please stick around and join us for lunch.

My father rose and limped directly to the door. We left without speaking to anyone. The first half hour of the drive back we were quiet. Over the radio came news of the lengthening drought. The governor, according to the voice of the newscaster, had just declared a drought emergency for all of southwest Kansas. The federal government had also designated it as a drought disaster area. These declarations meant farmers could pump more water and get more loans. Later I learned that such droughts were increasing in southwest Kansas, part of an unprecedented megadrought that had gripped the southern Plains over the last decade.

I took back roads into a north wind past miles of cornfields and empty farmhouses and the thinly disguised hog barns and a new flat-topped dairy. Right across from the dairy I noticed a stack of alfalfa bales the size of a castle or a warehouse. The cube they formed was so big it made the forklifts and trailers look like toys.

What's all that?

Milk cows need good hay. Lots of protein.

Do they get it here?

Yeah. Heard guys are switching to it. We'll have to ask the Hay Guy.

The wind whistled and pushed against the truck.

What did you think about all that?

Oh my god, I can't stand the doubletalk, my father said. It's like some of their mouths are full of marshmallows.

I laughed. I know it.

Nobody, except maybe the guy at Water Resources, said a word about what's really going on. They're draining the water as fast as they can and they're gonna just use it all up till it is gone, he said.

Gotta keep doing what we're doing, I said.

We gotta do more of it and then we'll just pipe more in and it's fine.

They must think everybody's dumb.

Marshmallow talk, he said.

I agreed with my father about that and I knew he agreed with me too. We did not agree on everything but we both felt the same way about this. It was a good feeling.

We gotta get back and check on those calves.

Yeah.

Mick is coming by tomorrow.

Oh yeah?

He ain't staying but we'll get some pizzas or something.

That evening, as my father went through the routine accompanied by his current wife and adopted daughter, I noticed a peculiar cloudbank forming along the southwestern horizon. Something about its shape, gauzy and ominous at once, seemed familiar but I couldn't say why or what it called to mind. When the chores were finished, the family headed back to the house. I stood at the fence stirred by the clouds and so I walked down

FIGURE 11

the band of shortgrass between the Big Circle and the Cimarron, orbited by dogs.

That prairie was always my favorite place to walk. The corridor had never been plowed. I remembered it carpeted by buffalo grass and dotted by prickly pear and transected by paths that glittered with waterworn rocks. Now my boots crunched through gray stalks and withered cactus and the brown globes of gourds. Only little islands of the short curly grass remained. I slowed down and looked more closely at the ground I covered and it was all the same. I didn't know what had happened but supposed that only a massive dose of chemical spray could have caused it. There was no way to know if the poison was drift off the Big Circle or if a pilot dumped his overload there like my father said they sometimes did in the sage hills and I started to think about how those pilots must imagine the land but then my eye caught on something light and I bent over a small chert scraper. Percussion waves were visible on the shard of gray rock. The puckers and sworls were like fingerprints and the point snubbed flat. It fit perfectly into the curve of my palm. I thought about the hands that made it and recalled that I had found countless flakes of flint in this spot as a child and wondered if it may have been a camp and I tried to picture the people who inhabited it recently or thousands of years ago and what the stones flensed and, oddly enough, those enigmatic beings seemed somehow more tangible to me than the toxins leached into the grass or the spray plane itself. The longer I crouched amid the dead plants the more it was possible to imagine myself as a wanderer trying to interpret the future rubble of our society. Uncertain, I got up and kept walking over the two low humps that were all that remained of the Beatty Ranch and past a line of fence posts still buried almost to the top by dust storms and when I crested the riverbank a mile farther I stopped. Below me the dogs were digging up the burrows of desert mice. A hawk hovered low, tail twisting against the wind. As I sat on the bank overlooking the sand hills, the shadows lengthened and in the softening light the mounds grew indistinct and slowly merged together into one mass of land. Gradually a vague sense of apprehension grew as if I were waiting for a message that I was not certain would arrive. I listened but heard only distant motors droning in the breeze. My gaze drifted from the sage-scented slope up to the lip of the sky. The cloudbank had taken new form. The billows of moisture caught the last of the western light and the vapors glowed like smoke from a fire. I recalled that years before in the Gran Chaco, a far-off place which now seemed like a dream, I had been surrounded by smoke from burning forests that resembled exactly these shapes towering one on top of the other. The flaming trees glowed in just this shade of ember red and were just as

difficult to comprehend. For years I was troubled by the feeling
that there was something inexplicably familiar about the
smoke from the devastation of the Chaco forest but it was only
then, sitting on the bank of the dry river near the Little Rock
House, that I realized the smoke must have reminded me of
Plains sunsets and that now those sunset clouds reminded me
of burning forests and there was no longer any linear causality
between the two images. As I shifted my weight, I suddenly felt
the scraper in my back pocket. I had no memory of taking it. I
was certain that I did not consciously decide to do so. But
there it was. Its unexpected reappearance startled me. I could
not decide if this was the message I was waiting for. If so it did
not seem to bode well. I was certain that I should not have
moved the stone. To return it was the least I could do. I got to
my feet and tried to retrace my steps. I walked a mile back and
circled round and round in the deepening dark. I could not
find the spot. When I thought I got close I laid the scraper
down. It was impossible to know if I returned it to its rightful
place or if it would stay there.

* * *

The next afternoon my father's brother Mick and his wife arrived in a blue truck they had driven from upstate New York
where he'd recently retired from his job as a grazing consultant
in the cooperative extension office that served the Hudson
Valley. A far cry from the kind of farming he'd learned on these
Plains, it was mostly telling trust-funders how to start and run
small operations. He still looked like a Plainsman in his weath-
ered jeans and the shoulders that tackled full grown steers were
still broad. We walked out to look over the red yearlings and
calves. My father said he needed to work them but had no one

to help. Mick looked at him. You got one son and he's standing right here.

We went back to the house and sat around the kitchen table to eat pepperoni pizza off paper plates. Mick was a masterful storyteller if you could get him started, but they weren't staying so the visit would be quick. I told him I was going through Fern's files and realized I didn't know much about her.

Well, he said. You know early in her life she was always wanting to run off, go somewhere, leave us, do something else. I don't think she ever wanted to be a mother. My father and Mick's wife were listening quietly.

When I was a kid she was cooking something on the stove. She hated cooking for all the men, you know. I was standing there playing and reaching up like I was going to touch the pan. I was three or four years old. I guess it's one of my earliest memories. She was telling me to stop it, just quit it, she said, and she'd slap my hand away and when she did that I must have thought it was a fun game because I kept reaching up for the pan. Finally she said you want to know how it feels? Do you? Then she goes and grabs my hand and put it on the burner. And held it on there. That's what you get. That stove burned off all the fingertips on my left hand.

Mick turned his thick hand over so we could see the glossy tips. My father looked over at him and said I didn't know that.

I guess you weren't around then.

Nobody told me about it.

I don't know why they would.

I guess she calmed down later.

I suppose you could say that.

We sat around the table and chewed the pizza. As the evening faded into night Mick poured tea into his plastic cup and then he went on to tell us more about Fern and the time she lost her mind. Her breakdown happened when I was about thirteen or fourteen years old, he resumed after a stretch of small talk. We were at church on a Sunday. Everybody in the family was there. I wasn't in the main room and I don't know where I was but I suppose I was off in another room having a Bible class or something. Well we heard a bunch of noise and squalling and I went to see what was going on. I got out there and here they were dragging Fern out of the building. Two people, one on each side of her, holding her by the arms, taking her outside. I remember it was Helen and someone else. Fern was bawling and blabbering. What is going on? When I asked her she was walking past me with those two people and I'll never forget how she looked over at me and didn't stop walking and she looked right at me and said it doesn't concern you at all. Stay out of it. And that was it.

Gol lee, my father said.

You don't remember any of that?

Nope.

You sure? None of it?

None of it.

Did you hear anything?

Not really. Little things I guess. Who told you what happened?

Nobody told me anything. I only found out by kind of piecing it together and I never did find out the whole story. None of the family would talk about it directly and nobody else would either. People talked around it. Maybe that's why you don't remember any of it, it's like they all wanted us to forget the whole thing ever happened. And who knows, if I hadn't heard all the noise that Fern made and come out and looked her in the eye, maybe I would have just gone along with it too. I know RW didn't want anybody to talk about it and that old man was the one who built the damn church and he led us all around like hogs with rings in our noses and Roy lost face with RW and the church over it so he didn't want to talk about it either. The closest I came to the full story was that she stood up in front of the church and called them all hypocrites. I don't know if they tried to shame her about something with someone's husband or something. But she apparently yelled at them for judging her when they were all sinners too. Women just didn't do that back then. She was probably right. But what they said was that she had a breakdown, she went nuts, she was crazy, you know. After that episode they just sent her away. Now I know you must remember that.

Not really. I knew she went away but that was it.

You didn't hear anything else?

Maybe but I don't recall it.

What were you doing back then?

Other things I guess.

I believe, Mick continued after a long pause, that first they sent her up to the mental hospital in Larned. Roy took me up to visit her in the hospital. We only went that one time that I can recall but maybe we went twice. You remember that because you were there, too.

No I wasn't.

You didn't go with us?

No.

I swear I thought you were with us.

No, I would have remembered that.

Maybe it was just me and Roy although I don't know why that would be. Anyway we went up there and saw her. I remember going through the doors with the locks and the windows with the bars on them. But we got into her wing and went into her room and there she was. She was just like herself, normal and I guess you could say acidic like always, just in a hospital room with the clothes and all. I wouldn't say she was happy to see us but we visited for a while and I wish I could remember what she said. On our way out they told Roy that whatever they were doing wasn't working so then they sent her off to a private clinic in Wichita for a new treatment they called shock therapy. She was in Larned for a couple of months and not very long in Wichita. Maybe a couple weeks or a month was all. She couldn't remember much about it except her fear of going into that room and how much it hurt. But when she came out, man, she was like a zombie. And full of drugs. Pills to sleep, pills to get going in the morning. Then the sleeping pills quit working so good and she figured out they worked better with booze. She was out of it every night. Just incoherent, you know. Rambling and jabbering and not making any sense. You really don't remember any of that?

No, not really.

Oh come on.

I don't know why I can't remember any of that.

Well. I'd sit and talk with her until she fell asleep. Every night I'd sit there just to make sure she didn't fall over or catch something on fire with a cigarette. If she didn't have her pills she'd stay awake all night long just talking and talking. All of it nonsense. After a couple of years she was less like a zombie and she slowly started to come out of it but she was never the same as before. It was like things got shook loose and she was always trying to put them back.

Thinking back on it now, he resumed after a pause, it seems like it was all that work she did on Wagonbed Springs that got her lined out again. It gave her something to do. She worked hard at it. It got to where she saw that as her life's purpose. She wanted to get the marker put down there. She always said it wasn't in the right place and it just had to be in the right place or else it didn't count. She won that battle. They moved it and made a deal out if it and she tried to become a respectable part of the community again and I believe she even came forward at Church and then she died not too long after that.

Mick was so immersed in his story and I in listening to him that I did not notice how late it had grown. The orange grease from the pizza had congealed on the paper plates and the tea jug was almost empty. Mick's wife, who had been patiently listening, put a hand on his leg. They had to get on the road.

After they said goodbye my father went to bed without comment. I walked back to the room in the old barn, which my father had since fitted with a bed, a shower, and an oak table I made in high school. I sat up late writing down as much as I could recall of what Mick told us that evening.

Above the table where I wrote hung a large black-and-white photograph, perhaps three feet wide and presumably taken by government surveyors in 1939. The image showed the earliest aerial view of the Cimarron and the Lower Springs. As far as I know it is the only print of this picture that exists. It looks like a medical scan of an internal organ, private and clinical at once. In the photograph, the already diminished trail of water appears as a dark braid through serpentine curves of pale sand. Save one field in the uplands, the land is blanketed with shortgrass. Dozens of arroyos and gullies are traced like veins across the hillsides. Wagon ruts cut across them and converge like sunbeams onto a single spot. This marks the site of the Springs.

When I paused to try to remember one detail of Mick's story, I looked up at the rare photograph, which hung for years in Fern's office, and I recalled how the picture appeared to me as a child, when it was positioned on the wall behind her black button-key typewriter. The longer I examined the image that

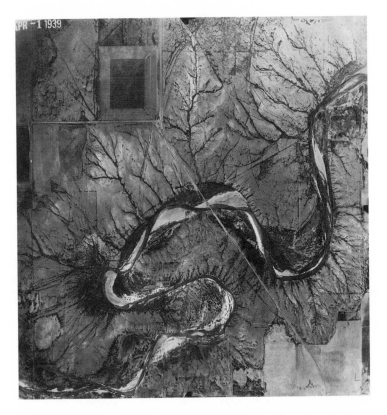

FIGURE 12

night, the more mysterious it became. Gradually it reappeared to me as if it was cloaked by the incomplete revelations about Fern. The tracks seemed like scars. I could not stop trying to imagine how the photograph appeared to Fern when she emerged from the hospital. I wondered what she might have seen in its visceral lines.

Fern left no account of the hospital. A journalist named Betty Wells did. In a six-part exposé printed in the Wichita newspaper in 1974, Wells described her own eight-day stay at the same hospital where Fern was first sent. Getting admitted, she noted, was surprisingly easy. Wells called and said she just didn't seem to be herself anymore. The building looked like an old brick house on a college campus. She signed papers in a barren lobby and was interviewed by a doctor. He focused on her sleeping and sexual habits. He asked her to say what crossed her mind when she heard clichés such as you can lead a horse to water but you can't make him drink.

The doctor admitted her with a diagnosis of reactive depression. Once inside, she recalled, "time seemed to melt together." For the thirty-five women patients, days began at 5:45 a.m. Life was organized around four doses of medication, at 8:00 a.m., 11:00 a.m., 4:00 p.m., and 8:00 p.m. In between the women did chores, played cards, and watched mindless television programs. "I learned," she wrote, "that life in a mental institution is based on waiting."[60] They gave her drugs in a little cup of pink liquid. She thought they would not overpower her mind. "By the third day of medication," she wrote, "I was a wreck." After the morning doses, Wells felt like she could do anything. Move a mountain or run a four-minute mile. "By early afternoon I'd be dragging."[61] The afternoon dose caused a letdown and none of the women could stay awake after the evening medicine. Her fellow patients ranged from teenagers to lifelong inmates. Some were confined naked in isolation cells or strapped to chairs. One suicidal mother of three said she just couldn't cope with her environment. Another had stabbed her abusive husband. Joan, an eighteen-year-old, was sentenced to a month and a half after she pulled a fire alarm at a county courthouse. When Wells was discharged, the doctor gave her four bottles of pills to take home: two antipsychotics, one tranquilizer, and one anticholinergic to calm her drug-induced tremors. Afterwards she wrote that "more than anything else, the experience made her feel dehumanized."[62]

Was it possible that Fern managed to survive two months in the Larned hospital with her will intact? If so, this now seemed a feat of tremendous strength and resolve. I wanted to believe it. Electroconvulsive therapy (ECT) left no room for resistance. This treatment used electrodes attached at the head to send high-voltage currents through certain regions of a patient's brain in order to induce seizures. The seizures were so violent that patients, even when restrained, would often fracture their own bones or shatter their teeth.[63] Kansas doctors heralded this treatment as a breakthrough.[64] Women made up a disproportionate number of those treated: more than two-thirds of the patients nationwide. Electroshock targeted a range of disorders, including behaviors that violated gender norms. In the late 1950s, state hospital patients could be given electroshock without consent and in some places husbands could sign up their wives. The almost universal effect of ECT was amnesia. Common memory disturbances included forgetting family bonds or husbands or children, forgetting everyday routines, and forgetting the major events of the present or the past. In fact, this amnesia was so pronounced that many women subjected to electroshock treatments presumed the side-effect of memory loss was its intended purpose.[65]

I recalled Fern's dentures and I wondered if memory failed me too. For a few days after Mick's visit I went around like a sleepwalker, unable to stop myself from picturing what this episode meant for my grandmother's life. I struggled to make sense of the light it now cast back on her obsessive pursuit of the Springs and her fragmentary files. In its glare, the men who committed her appeared monstrous. Try as I might, I could never fathom the depths of what happened. But surface resonances with the aquifer haunted me. She was sent to the hospital around the same time her father's wells drained the Springs.

I can only guess whether she ever pondered how the fall of the aquifer and the rise of irrigation seemed fused with the vicissitudes of her own life. All I know is that she was undoubtedly aware that deep-well irrigation caused Wagonbed Springs to go dry. She knew that her father was responsible for drilling the wells nearest the Springs and for institutionalizing her. And she spent her post-hospital life fighting to relocate the memory of those vanished waters.

As the months passed, something of her search began to seep into my own. I realized I was still searching for a deeper kind of sustenance in the aquifer waters, too. Something about their destruction and the resulting absences was bringing me closer to the elements that made up my inheritance. They raised questions I hadn't expected and didn't know how to understand, even as they revealed dark holes in what I thought I knew about the family ties I had taken for granted. They made me realize that I could never really put myself in Fern's place. But, like her, I kept assembling fragments. Like her, I must have hoped they would add up to something coherent. And perhaps like her, I struggled to make the pieces fit.

* * *

The pale sand stretched to the blue horizon, scraped to brown hardpan between yellow dunes. The circle of sand was all that was left after the water had run out and farmers abandoned the field. The land's prior contours were traced by fence posts that now floated in the air or were entirely submerged. Previously buried gas lines for irrigation motors lay on the surface, flat and ominous like black snakes in the sun. It looked like the floor of a giant hourglass after time had run out.

The sand drifted across the two-track. The truck bucked and bogged as my father gunned it. Beside him sat Mike, the Hay Guy. He cut tens of thousands of bales each year, so much that the plants wore through the metal plates on his machines. We found him in his barn replacing a swather platform. My father had called to see if we could talk to him about hay and water. The Hay Guy said he could do better—he'd show us some shit we wouldn't believe.

So we headed east into the sand hills on the north end of the county. Mike saw the dunes when he was involved with the state Conservation Reserve Enhancement Program (CREP). Started in 2007, CREP paid farmers along the upper Arkansas basin to take irrigated land out of production for fifteen years, retire the water rights, and establish native grass cover on the acreage.[66] Mike knew those who helped to replant the grass. We passed miles of open dunes. We listened to him narrate each circle of sand.

He told us how the blowsand was created. Until the 1970s, agriculture was mostly confined to the flat, rich uplands. The sand hills defied cultivation. It wasn't until pivot systems could travel over the bumps that they decided to plow up or spray off the sagebrush and try to make the sand hills grow crops. By 2007, it was clear that agriculture in the sand hills was unsustainable. Without the sagebrush to hold down the soil and without water to grow new plants, the sands would begin to blow.

To address the problem, the CREP prescribed a three-year transition from irrigated crops to grass. The first year, farmers who participated were supposed to water a cover crop. The following two, participants watered grass seed to get it established. The fourth year, farmers were supposed to turn off the sprinklers. In return for participating, CREP gave farmers a sign-up bonus and annual payments. There was an option for enrolled farmers to take the full CREP payments up front for a 20 percent reduction of the total payout.[67] Many landowners followed the rules. But some farmers signed up for the program, took the payment, and didn't plant the grass. Some didn't want to spend the money it would take to establish a proper cover crop. Others signed up for CREP money only when they had already run out of water. Those people used the water until it was gone, then got paid to enroll in a conservation program, then didn't have enough water left to actually transition to sustainable grass. Without solid cover, the sand shifted and blew.

Mike knew who owned each field and who farmed it. I was surprised how many were owned by companies. The first four sand circles we passed belonged to a big alfalfa-growing corporation. It was linked to one of the many megadairies that moved into southwest Kansas during the prior decade's dairy boom. Megadairies are more like factories than farms. They have automated systems that milk thousands of cows, process the milk out of state, and are usually part of large conglomerates that operate across regions. Local development groups in southwest Kansas actively courted dairy companies that had run into regulatory issues elsewhere, hosted dairy days, and offered them various financial incentives to relocate. This meant that while dairy numbers declined nationwide, they sharply increased in southwest Kansas from 2007 to 2017. Industry proclaimed the area "Dairy's New Frontier."[68]

The megadairies increased pressure on the already strained aquifer. Dairy cows need to eat high-protein alfalfa hay if they are to produce industrial quantities of milk. Alfalfa plants contain the most protein and calcium in the early stages of growth. That means dairy hay is cut every twenty-five to thirty days, creates smaller yields for the farmer, and requires the greatest input of fertilizer and herbicide. It also takes a lot of water to grow. In western Kansas, alfalfa plants require nearly twice as much irrigation as corn.[69] The corporate dairy demand for alfalfa has caused some farmers to switch from corn to alfalfa to meet this market. Mike knew most of the local alfalfa growers. One was a man I'll call Reilly. We made a detour to say hi.

Reilly was on a tractor, trying to drill a game-mix of grass seeds into the blowsand. The lugged wheels lurched and spun. We watched from the truck. When he got to a firm spot, he idled the motor and came over to sit with us in the air-conditioned cab. He said he didn't think the seed would grow but what the hell. Reilly farmed eight quarter-sections of alfalfa. That year he sold six thousand bales at 100 dollars each, along with eighteen thousand tons of silage. His major customers were megadairies. He told us how they had added calf operations to their milking lots and brought just-weaned heifers from Michigan to feed on aquifer-watered alfalfa and corn. One dairy alone was planning to expand its local operation by twenty thousand head the next year.

Reilly told us he was uncertain of the future and sometimes wondered if he had been tricked by the company. Alfalfa farmers in the area usually do not sell to a cooperative or elevator. They contract directly with the buyer. When the dairy started several years ago, it bought his hay at 205 dollars per ton. By 2017, it was paying only around half. He ran through a dizzying set of figures about prices, yields, and acres. The big difference, he explained, was the corporate investment in farmland. The same dairy had purchased 255 quarter-section fields in southwest Kansas and 80 in nearby southeast Colorado. That is more than fifty-three thousand irrigated acres. Some were converted from corn to alfalfa. The company then flooded the local market with its own hay. This drove down farmers' prices until growers like Reilly were barely breaking even. But the corporation, he noted, was doing good business. In fact, the value of alfalfa hay grown with aquifer water was now so low that some people suspected the company bought local hay at a fraction of its market value, shipped it out on its trucks, and resold it for a profit elsewhere.

If true, then corporate profits are a key part of the aquifer depletion puzzle. It should have come as no surprise. The scale of industrial farming is staggering. Southwest Kansas is home to some of the nation's largest corporate feeders, beef- and poultry-packing plants, slaughterhouses, dairies, milk-drying plants, and hog farms. More than 2.5 million beef cattle live there in feedlots that handle tens of thousands of animals.[70] Just across the Oklahoma line, one company processes 5.6 million hogs per year in its plant. A subsidiary firm makes 55 million gallons of biodiesel each year from its pork fat alone.[71] Another company runs three corn ethanol plants with a capacity of 205 million gallons per year and sells the by-products as feed.[72] Multinational meat-packing companies operate slaughterhouses that process several thousand cattle each day.[73] All are billion-dollar businesses. They drive farmers' choices to produce corn, silage, sorghum, or alfalfa. Their profits depend on aquifer depletion. In other words, there is a multibillion-dollar corporate interest to prevent regulation and to pump the water until it's gone.

Kansas groundwater users are required to self-report their yearly water usage. In 2018, investigative reporter Karen Dillon used an open records request to access these reports. She compiled a list of the top water users and calculated how much water they extracted from 2005 to 2017. She found that the top 2 percent of water users consumed 22 percent of the groundwater that was pumped statewide. Moreover, she found that the top users of the aquifer were these agribusiness corporations and the tenant farmers who leased their land or farmed for shares at their direction.[74]

In the days after our conversation with Mike and Reilly, I finally began to see the fields and factories as parts of a single industrial floor. But the scope of corporate influence is largely hidden from the Plains public. One of the places it hides is behind the fiction of a family farm that is opposed to industry and elitism. Depletion, we are told, rests in the hands of family farmers whose choices are part of an independent livelihood imbued with moral values. These values are the foundations of a community based on hard work, equal opportunities, and taking care of neighbors. The more I learned about corporate dependency on aquifer use, the more this seemed like a trick. Making local farmers appear solely responsible for depletion lets industry and distant landowners off the hook. It obfuscates the complicated ties that link depletion to the financial operations of farmers, banks, government programs, and corporate profits. And it smears any critique of overuse as an attack on community values or small farmers. But the closer I looked, the harder it was to distinguish between family and factory farms. The line was more than unclear. It was intentionally blurred.

Like other midwestern states, Kansas ostensibly prevents corporations from owning agricultural lands by law. Created in 1931, the law was intended to protect local farms from the big corporate interests that laid the foundations for the Dust Bowl. Since then, the law has been steadily weakened. Its present version permits certain kinds of corporations to own land, including family farm corporations, authorized farm corporations, limited liability agricultural companies, family trusts, and authorized trusts. Moreover, it allows feedlots, dairies, and corporate hog operations to acquire agricultural land and to appeal county efforts to oppose their expansion.[75] This empowers agribusiness corporations. It also means that it is very difficult to tell how much land is owned by authorized corporations or shell LLCs, and how much land is owned by local residents or family farmers. But there seems to be a serious imbalance. Absentee landowners, corporate or not, are responsible for a major portion of aquifer use. In the county around the Little Rock House, at least 60 percent of the farmland is owned by nonresidents.[76]

As we drove past miles of bare sand, I assumed that Mike and Reilly shared my sense of who was responsible for the devastation. I was wrong. There was no ready critique of corporations or elites or suitcase farmers to be heard. Reilly blamed local farmers for not adopting more efficient technology. He scrolled through his phone and showed me videos of sprinkler nozzles working in the wind. He was looking to buy land with better water so he could keep pumping from the aquifer to grow hay. I said something critical. Mike jumped to defend those who wrought the destruction he had shown us.

People have the right to do what they want with their land, he said, you can't tell them what to do. You know, he continued after a long pause, they say there are two kinds of people. Smart people and educated people. Me, I'm not educated. I don't know a lot. But educated people don't know shit.

My father chuckled.

It's easy to sit back and criticize this or that from a distance, he continued, but what would you do? Probably screw it up even worse.

The road was bad and getting worse. The sand was deep. Even in four-wheel drive, the truck was dragging. We paused on a high spot and surveyed a long sandy stretch to come. My father decided to turn around. We'd seen enough anyways. He

FIGURE 13

drove back past the same dunes of sand, heading towards Mike's house. I asked about water.

The Hay Guy said he knew the water was running out but there was no need to regulate it. Hell, the economics will sort out the problem better than the government or the experts can, he said. It just ain't feasible for a lot of people to keep going. Farmers aren't stupid. They are businessmen.

I'd heard people say this my whole life. But as I stared at the barren dunes, the dissonance between the idea and the reality of the wrecked fields seemed to join with a faint feeling of motion sickness caused by the pitching truck. What was Mike defensive about, if not the sense that farmers' choices were somehow limited, that it wasn't really their fault? But how did that coexist with the myths of responsible industry and the farmer-as-businessman and the recurring devastation of the land all at once? At the time, I felt those questions only as stirrings of nausea layered under the scenes of blowsand yawing past the window of the truck. As I recall those scenes and the seasickness that accompanied them, I find both superimposed with moving pictures from the Dust Bowl era that I had seen years before and which appeared again, strangely distorted. The filmic and photographic images, which have since merged into a single montage in my mind, were taken near the same spot where we drove, and depicted what could be mistaken for the very same expanses of scoured hardpan and drifts piled around a vacant house exactly like the one just north of the Little Rock House in which Fern spent her Dust Bowl childhood.

I found nothing in Fern's files about the Dust Bowl. Nowhere is there anything to tell me what she saw or how she experienced one of the nation's worst ecological disasters, whose decade-long upheavals were the backdrop to her most formative years. She was six when the dust storms started and fifteen when the droughts eased. Perhaps more revealing is that her archives do not include a folder about that time at all. She had files for crops, social clubs, banks, sports, churches, feedlots, and miscellany, but not one for this calamitous series of crashes and droughts and dust storms. The absence was striking. There was no way to know if her avoidance of the Dirty Thirties was intentional or if she did not register its eventfulness or if portions of her memory were erased or for some other reason entirely. When I try to remember what she told me about it, I recall her referring to those years only a handful of times and even then my memory retains only the sparest of childlike details. Those details provided my first introduction to the Dust Bowl, a period so familiar it could easily be taken for granted.

It remains stuck in my mind, without any context or evidence whatsoever, that she described sleeping with a wet rag over her face and recounted dust storms that blotted out all visibility for days and that dirt sifted through the frame of the house to make miniature drifts on the sills, to her mother's lasting annoyance. I believe that she must have told me those things but I am no longer certain. Fern made only a single comment on the Dust Bowl in all of her later historical work. I stumbled across it by accident in the State Historical Society Archives, where I happened to discover transcripts of the 1972 radio program *Glimpses of Grant County* that Fern wrote and which aired in serial form on the local AM station.

"It is hoped," she wrote in a preface to the transcript, "that readers will forgive repetition and over-simplification, bearing in mind that the original presentation as radio broadcasts required each Glimpse to be complete within itself and limited to one minute." The episode she titled "Dust Bowl Daze" followed a piece on the building boom and came before one on highways. The full script, which I can almost hear spoken aloud in her deliberate smoker's voice over the staticky radio waves, is as follows. "Some residents nearly panicked in the spring of 1953," she began, "when several days of blowing dust darkened the atmosphere and tumbleweeds stacked up to the second story windows at the hotel. But that was just a prolonged high breeze to us veterans of the Dirty Thirties. Tales have grown pretty tall about those times," she said, "and about the only good things to come from them was the westerner's determination to never let it happen again and a certain grim humor. One 1935 story told by the theater owner," she continued, "was that the best place to see the movie was in the projection booth, because the picture reflected on the dust in the air before it could reach the screen. And Roy H., manager of Miracle grocery then, actually planted a garden in the dirt that accumulated in the store window, with radishes, beans and onions thriving in the good rich topsoil. Might as well joke about the Dust Bowl," she concluded, "because if you weren't here then, you wouldn't believe it anyway."

I imagine a ten-year-old Fern huddled in the darkness of the theater, witness to the peculiar transposition of picture and dust in the beams of the projector. I wonder if she and her fellow moviegoers were frustrated by the dust interrupting the coherence of the screen, or if any of them were instead amazed at the way the images transformed the particles of dust into thousands of tiny pictures that floated in the air and settled on their bodies, literally up for grabs.

Surely the town cinemas were one of the only spaces of refuge from the overwhelming magnitude of the crisis. Perhaps others in the cinemas sought to escape the polarized interpretations of that moment. Almost immediately, the deepest American myths were projected onto the blowing dust. Coastal commentator H. L. Mencken invoked it to argue that Plains residents were biologically inferior, uncivilized, and best fit for sterilization. Local chambers of commerce formed so-called truth squads to gloss over the extent of the problem and promote the region's business interests. Artists and socialist intellectuals decried the Plains as the site where dreams of agrarian equality were exhausted into a wasteland of corruption and greed.[77] Others called it a test to be overcome with fortitude, ingenuity, and Providence. Then, as now, these opposing myths set the polarized terms by which environmental loss was understood.

* * *

In 1940–41, anthropologist Earl Bell carried out four months of ethnographic fieldwork in southwest Kansas. He was later joined by Irving Rusinow, a government photographer. The team was tasked with recording rural life in Haskell County, which sits just

FIGURE 14

east of the Little Rock House, as part of the USDA's Bureau of
Agricultural Economics comparative study of six rural commu-
nities across the United States.[78] The county was already iconic.
Bell's work followed up on a 1939 report by sociologist A. D.
Edwards, based on three months of fieldwork, that declared
Haskell County an exemplary community of the Dust Bowl re-
gion. Edwards examined the influence of drought.[79] Bell docu-
mented the social attitudes linked to the devastation.[80]

Bell and Edwards concurred on the general pattern of the
ruin, the economic and ecological collapse, the cinematic dust
storms, the exodus. They linked it to the near total depopula-
tion of the area during a prior drought between 1893 and 1900.[81]
They identified as culprits the newcomers attracted by the
wheat boom of the 1910s and 1920s, when the remaining prairie
was plowed under, 94 percent of the land was planted to wheat,

and the farmers prospered. In 1930, more than half of the resi-
dent farmers had lived in the area less than ten years.[82] This
second wave of settlers, they agreed, laid the foundations for
disaster. These included extractive cash-crop monoculture, ad-
herence to arcane theories of soil and moisture, techniques
such as dust mulching, rapid adoption of machine farming, and
a dramatic increase in nonresident landowners, who in 1936
owned more than half of the county. The major cause of the
crisis, they agreed, was the dominant profiteering attitudes and
the belief that farming was the same as business. "The Haskell
County farmer," wrote Bell, "believes himself as much a businessman as a manufacturer is."[83]

If so, their businesses failed. A glut of wheat dropped prices
from one dollar per bushel in 1929 to sixty-three cents per
bushel in 1930. To make up the difference, Haskell County
farmers planted more acres to wheat than ever before. The 1931
harvest was the largest to date. The price of wheat dropped to
twenty-five cents per bushel that year and twenty-three cents
the year after. Severe drought began in 1932 and lasted several
years. Farmers kept using techniques unsuited for arid climates.
Without rain, nothing grew. Absentee landowners sold and
bought. New residents left. The dust blew in the black blizzards
of 1933, 1934, 1935, 1936, 1937, and 1938. It was in the aftermath of
the huge storm on April 14, 1935, that an eastern journalist first
described this area as the dust bowl of the continent. The dust
caused fevers and pneumonia and suicide and madness.[84] According to Bell, second-wave settlers could not cope with the
violent fluctuations of the Plains. "Each new uncertainty," he
wrote, "left its share of scar tissue in the gnarly form of the local
social structure." If Bell and Edwards are to be believed, a distinct kind of settler society coalesced among those who
stayed.[85]

FIGURE 15

It was based on federal money and the search for water. The disaster forever tied Plains agriculture to government aid. It poured into Haskell County through various programs. Residents called it a lifesaver. One-fourth of the county's populace was on direct welfare between 1932 and 1936. More than three hundred thousand dollars' worth of low-interest loans were made. The federal Works Progress Administration projects employed 10 percent of the county's workers. Most of the money went to farmers through the Agricultural Adjustment Administration program (AAA), which paid famers for failed crops and to reduce plowed acreage. AAA payments to Haskell County averaged more than $450,000 per year from 1933 to 1936. They assured a fixed income whether a crop was harvested or not. In 1936, 90 percent of farm operators in the county received benefit payments. However, the amount of the payment was based on the average of each farmer's harvests in 1929, 1930, and 1931.[86] In other words, those who were the most responsible for the Dust Bowl—the most unscrupulous and exploitative in the years prior—received more aid than those who had voluntarily conserved.[87] The dependency grew. In 1939, government payments accounted for approximately half of the total agricultural income of the county. Some residents called it farming the AAA.[88]

Aid, industrial farming, and environmental collapse eroded social ties. At the end of the 1930s, Bell found little that resembled a community at all. Instead, he described a set of individualistic people with cosmopolitan tastes, strong capitalist ambitions, and weak ties to neighbors and to the land, which they still viewed not as a home but as the site of industry. Southwest Kansas farmers, he wrote, did not imagine they belonged to a single village. Instead, they drove around an area some hundred miles in diameter seeking to increase their profits.[89] Alongside this not entirely rural pattern, Edwards and Bell were struck by what they called a "social psychology peculiar to the area."[90] It was marked by speculative ambition, willingness to gamble, and a fanciful optimism that they found nearly unbelievable.[91] "The faith in luck is reflected in their entire personality organization," Bell wrote, "and is indicative of their inability to develop a method of agriculture adapted to the environment." Photographs of grain piles hung in nearly every living room. "The farmers' main recreation," Bell noted, "is going to the cinema."[92]

The aquifer tied Dust Bowl attitudes to groundwater depletion. Southwest GMD now covers many of the same lands ruined by 1930s agriculture. Those who stayed were singularly focused on groundwater. "This is good country," an old man told Bell in 1940. "All it needs is water and it will produce better than any land in the world."[93] By the end of the 1930s, water became fused with hopes of all kinds: for prosperity, for blessings, for love, for family, for triumph, for order, for reason, for civilization, for freedom, for willpower, for salvation. Only water, one Grant County farmer said in 1941, could erase what he called the original sin of industrial agriculture. "We know our land is still a garden spot if there's water."[94]

People back then started to think about the aquifer as underground rain. They said its bounty was limitless. They wrote how groundwater would "defy the rain gods" and imagined it held "more salvation than the heavens could pour down." They thought the inexhaustible flow was theirs for the taking.[95] "While pumping for irrigation might temporarily lower the groundwater table," a Topeka reporter wrote in 1927, "it does not appear probable that the groundwater supply can ever be exhausted."[96] Deep wells, newspapers reported, would make western Kansas "an Eden" once again.[97] Faith in the abundance of these waters put an end to the more sustainable farming techniques that were beginning to be adopted by the end of the 1930s, as well as the progressive policies that accompanied them. One historic opening was lost with them.

"The best defense against a recurrence of the Dust Bowl," local farmers told a reporter in 1947, "is irrigation from underground wells."[98] Farmers said they started to look down for their moisture instead of up. Sociologist William Mays revisited Haskell County in 1965 to update Bell's work. The biggest change since Bell's analysis, noted Mays, was the rise of irrigation. At the time, more than two hundred thousand acres in Haskell County were already being watered. Mays called irrigation a "stabilizer of agriculture."[99] Like the residents, he stated that irrigation offset uncertainties of the environment and market. By 1965, locals considered farming a rational economic enterprise. "The population," wrote Mays, "now accepts as their ideal-type the farmer-capitalist, or agri-businessman, who has won out against great odds."[100]

FIGURE 16

"Farm families go to the movies often," Bell wrote, "an average of two a week is not unusual."[101] There is nothing in Fern's notes to tell me what moving pictures she saw in the mid-1930s, as the dust storms raged and the people fled. Was it the 1936 government newsreel *The Plow that Broke the Plains*, which portrayed the folly of Plains farming and featured the images that could so easily be projected onto the blowsands of the present?[102] I wonder how she would have perceived the shots that showed her nearby surroundings flickering across the dusty screen, intoned by a narrative of ruin. Or did she see instead, as I supposed she must have, the Gene Autry film *Tumbling Tumbleweeds*, a story that ran with great success in Plains cinemas in 1935? It tells the story of the singing New Deal cowboy who returns home as part of a medicine show peddling a tincture called Dr. Parker's Painless Panacea to drought-stricken settlers. The white-hatted hero eventually solves the murder of his father at the hands of a corporate landowner due to a conflict over water rights for irrigation, rights that he presumably inherits along with their promise to return the land to moral fruition. The cowboy sings to crowds of townspeople dressed in clothes from the 1930s, and alongside horses and stagecoaches appear cars and airplanes and trucks. Elements from the present mingle with icons of the past to such a degree that causality is uncertain and time itself begins to erode. I do not know what Fern might have made of this peculiar shuffling of past and present. I don't know if she would have found depletion unjust or felt outrage at a rigged game. But the unresolved stories linger on the Plains, where they build like dust clouds on the horizon.

DUST

At dawn we headed south to Abilene to talk with Fern's brother Gene about water. My dad set it all up. He emailed Gene, said I was writing a book and that we wanted to interview him. Because my father asked, Gene and his wife Mary agreed.

Gene was in his mid-eighties at the time. When RW passed, he left the farm and a debt to his three children. Fern renounced the land and the burden. Gene and his other sister inherited both. Fifteen years after RW's death, they sold the land and broke even. My father bought back the Little Rock House. Gene and Mary retired to Abilene. She had family there. The town was also the southern Plains stronghold of the Church of Christ and home to Abilene Christian University. For them, the Church was almost the same as family.

Our first stop was for coffee and hash browns at the McDonald's in Shamrock. I ordered a latte.

Oh Lord, said the middle-aged cashier. I don't drink that stuff.

I know it, my father said. I tell him to just drink coffee like a regular person.

You know it is just espresso with milk, right?

I never tried it, the cashier said.

And I don't want to, added my father.

The temperature rose as we dropped from the tablelands. We passed towns with courthouse squares and false fronts while emptied houses fell down in arrested motion. Freshly plowed fields followed us the whole way.

We hit the dust cloud about ten miles south of Wellington. Dirt torn from bare earth draped a brown wall across the highway. The dust glittered in the sunlight. When we hit the cloud, the light dimmed. Particles hissed around the metal and glass. Visibility was reduced to a few meters on either side. I slowed the rocking truck to a crawl and turned on the flashers. I could no longer see the road ahead but as we pushed through the fog it broke and we were surrounded by thousands of phantom shapes. I watched the dusty spirals twist and begin to take a form I could recognize but in that exact instant the wisps would dissolve and just as quickly another indecipherable figure would swirl up and at once collapse. In the midst of the cloud it was impossible to tell if the earth was blowing away or regenerating itself or both at the same time. Against the rough asphalt the dust flattened into a sheet that rushed faster than our speed. The opposed motions of the mysterious spirals and the surface undertow and the lumbering truck seemed significant but before I could fully register their import we emerged from the dust cloud into a landscape that appeared somehow changed.

The sun burned red in a hazy sky over coppery fields. The terrain crumpled and smoothed and crumpled again around dry rivers with legendary names like the Mountain Fork of the Brazos. We passed the sites of homesteads and hide camps and massacres known and unknown. Mesquite clumps began to dot then overrun the unplowed margins. A group of feral hogs scattered as we passed.

It was fitting country for apocalyptic faith. The Church of Christ is one of the most conservative branches of the movement that aimed to restore first-century Christianity. Adherents take the Bible as their sole authority and charter for dominion over nature. They eschew instrumental music, alcohol, dancing, and swearing.[1] The militant message and colonial structure flourished on the west Texas frontier of Greer County, where the ancestors of Fern's mother settled in 1889. She carried the faith with her to Kansas. RW built the local Church of Christ for his wife. By the late 1980s, I was related to almost everyone in it. But the Church was not as inclusive as it seemed. RW told my father not to marry my mother, who was Catholic and thus an evildoer. My mother raised me and my sister as Catholics, too. My grandparents often took us to their church. Although people there were very kind to us, we always knew we were on the other side of some unspoken divide.

Wow. I haven't been down this road in forty or fifty years, he said. The last time I was going north a hundred miles an hour.

Oh yeah?

RW was desperate for me to go to Abilene Christian. So was Fern.

Huh.

She thought it would be a good influence. She thought I was too wild.

Didn't she call all the church people hypocrites?

Yep. I don't know what she was thinking. Preacher's kids. Worst people I ever met.

Really?

Kind of guys who would crap on your windshield when they knew you had a date.

I smiled.

Thou shalt not be demeaned by any worldly matters or urgings of the flesh when thou art traveling upon the high road to Abilene, I said with a twang. Whosoever hath filleth their cup from the kegs of Abilene shall know not suffering nor anguish for their righteousness shall be as their shepherd. The blessed waters art their rod and staff and shall comfort and keep from them the terrible thirst of the barrens and the fear of evil though they travel through the valley of the shadow of death.

He laughed. I realized how easy it came to him and how easy it had come to me. The uncomplicated moment made me realize that something had eased without me knowing it. I thought about the mysterious ways that trust and alliance could recharge almost imperceptibly between two people who started far apart. I listened to him joke about preachers and the sins of their children and the beauty queen he dated and how glad he was to leave after a year. After a stretch of small talk, I said I couldn't find anything Fern had published.

She wrote all the time but never published anything.

Did she try?

A little at first. But she was afraid to fail. Maybe that's why I went to Nashville to be a singer. I decided to just walk in and talk to people, give it a shot.

Oh yeah?

It almost worked. One guy was a big producer. Called me up, said Barbara is going to cut your song. It's gonna be a hit. He needed the rights. I didn't have them but said I could get them. They were tied up somewhere. Then your sister was born premature. Thought she was going to die. So it took me a while to get the rights transferred. Maybe four or five days. When I went back they said we don't want it anymore.

Huh.

So that was that. I don't regret it. Your sister was more important.

The story snagged like a hangnail in inflamed quick. Just like that, old defenses sprang up about blame and responsibility. I remember that I almost said something. But now I am glad that I didn't. Laughing with him was more important than pushing on scar tissue and I realized that he probably didn't mean it like that at all.

I fell silent until we hit Abilene. My father talked to his wife on the phone. We circled around campus and picked up some tacos. Something about the blocky clothes and diamonds and Whiteness made me uneasy.

The house was two blocks from the campus, on a quiet street of ranch homes. Theirs was blonde brick on brown Bermuda. Mary answered the door, in a pink top with a gold cross pendant. Her body was withered, but her honeyed voice and sharp eyes were just like I remembered. Gene shuffled up on unsteady legs. His hair was pure white and he had a hearing aid. His baritone voice made me feel like a child.

They greeted my father warmly. We sat on the couch in the tidy carpeted den. After the pleasantries were over, I asked if I could record. Mary sat in a rocking chair and said nothing. Gene looked at my father, not me. By then, we had done dozens of interviews together. He took the lead. My father told Gene I was writing a book about family history and irrigation and Fern.

I've been thinking about Fern lately, Gene began. She used to take care of me. But she had a wild hair and took off. I used to talk to her a little bit, you know. She'd look at me and say I can tell you got a little bit of rebellion in you too. She had more courage than I did. He paused. How are things up there?

Oh, that dirt is blowing today, my father said. You should see it. Everybody's got open ground.

It's dry up there?

Really dry. Big drought.

Now you want my history?

Sure, I said.

Gene said he was born on the Old Neese Place in 1932. Nothing remains to mark the spot. There used to be an old Model T there, he said, but the dirt storms buried it. I am old enough to remember the dust. It was so bad that one night we were eating our rabbit, that was all we ate back then, and the ceiling came down on us from all that dirt. Not long after that, Mama Faye moved us.

I bet, my father said.

When I started out we didn't have any irrigation wells at all. I tell you honestly, I wish we'd never had them, he said after a pause. We could have made it just as fine with dryland farming, conserving, rotation, summer fallowing, you know. We could've made just as much money. Pop got some royalty from gas wells. Well that spurred him. Maybe we need to try irrigation. The first wells, why, they was pretty good. Let's put down another one. We got some land over there. Let's put down another one.

He spoke slowly and deliberately, with long pauses between lines. Each story felt distilled or rehearsed. It was his voice that charged them with feeling. His inflections moved against the spare language like an undertow. I was surprised by the tones of bitterness that often underlay what he said.

The best well was down on the river just by your place. It was, oh, two thousand gallons a minute. Put in a ten-inch pipe and the water just shot out. When we first started if you got a well like that it was really something. The guys who drilled it came back and said it would be one of the best wells in southwest Kansas. The gravels it showed were so deep. So good. All the way down. Boy it was a good one. Now I understand the well on section 99, north of there, has no water. They had to stop using it. The others still have a little left. That's about it.

My father chuckled. That's a pretty fast summary of a life.

Somewhere a clock gonged the hour. Gene looked at me.

Are you married?

No.

What happened?

I don't know. I made mistakes. I move a lot.

Well, that's not smart.

I know it.

And what happened to your basketball team? Boy they lucked out getting in that tournament. Should've been Oklahoma State.

I don't disagree.

Your team hasn't won a game since Christmas. I don't know what's wrong. So what questions you got?

How about your dad, RW?

He was all for pumping. He didn't say let's do it. He said we will do it. Irrigation was so enticing and so available. In those days, cheap gas and good water, the county was booming. It was a paying proposition. And it would have been. Until Dad started feeding cattle. That was the downfall of our family. I hate to think about it. The rise and fall of the family. That's about it.

Mary laughed softly from her chair.

Anything else? my father asked.

To pay off that debt I had to sell just about everything, he said. My irrigation it was sweet and it was sour. Does that make sense? I don't know how many hours we spent. Leveling the land. Making ditches. Setting siphon tubes. Then we got pipe. Had to move it. We had two miles of aluminum pipe. I don't know whatever happened to all that pipe. It's too long ago.

Maybe the wind blew it away, my father said.

Maybe so. Are those old sprinkler heads still out there northeast of your house?

Yeah, I still got them. The scrap iron guys will give a hundred sixty-five a ton. I think I'll sell it.

Maybe with all that gold buried on the Cimarron you won't have to, said Gene.

That's right, I'm rich.

I don't know how else a man could make it out there.

Me neither.

Now you went to Abilene Christian too, right?

Yeah, said my father. Just one year.

Were you on the basketball team?

No I high jumped on the track team.

What was your best?

Six feet two.

Wow, how about that. I remember you were a pretty good basketball player in high school.

Gene didn't have much else to say about the farm and the water that was running out. But he knew the name of every boy who played on my father's high-school basketball team and the scores of games played in 1967. They talked about opposing teams and coaches and strategies. After a long while, he asked what else.

Well let me ask you, my father said, about that debt.

Gene first attributed the loss of the family farm to embezzlement. He recounted how a guy from the credit association stole three hundred thousand dollars by taking truckloads of corn and selling them on the side. Then he said that RW was to blame for cutting corners in the feedlot, which caused the creditors to call in loans that he couldn't repay. It all amounted to the same thing in the end. Father against son. The conflict was never resolved. Our relationship went pretty bad, Gene said. I hate to say it.

What happened after that?

It was already too late. Guys, I had over a million dollars of debt when RW died. You just can't make up a million dollars farming. He swiped at his eyes. The pig corporation came in and offered a premium price for land. I got enough to pay them all off. We had a little left. It was only through God's help. It hurt me. I did it. I didn't want to do it. But a million dollars of debt, how else do you pay it?

The stories came out polished, like time and repetition wore them down to a core. It was hard to imagine their original shape or hold them in place. Just when I thought I could grasp who was responsible for what, the plot dissolved. As I was sitting there, thinking about the ways the stories of depletion were always incomplete, their son Warren knocked on the door and walked into the room. A former preacher, he now worked for a missionary society.

You remember Luke? my father asked. He's writing a book. And it is not an unseemly book. At least so far. You wanna hear the kicker? He gets paid to write. He makes money by not working.

We all laughed. Warren sat down and the stories unraveled into loose threads about real work. They reminisced about herding sheep and shooting coyotes and fixing balers and laying out harvest dinners in the fields and the various hired hands they worked alongside.

I saw Ortega's widow the other day in town, my father said. He taught me to cowboy. How long did he work out there?

A couple three years, said Gene. Pretty good for a Mexican.

He looked over at me.

RW was kind of racist, Gene told me. He said there is one good Mexican in a thousand so we're about due for our first one.

Nobody laughed except him.

Oh my, Gene said. He looked at my father. You go to Church?

Nope, my father said. I haven't been in a long time. I don't like going in there.

Well. I hope you take care of your spiritual life. You're too special.

Send me a card on my birthday then. Luke, anything else?

No, that's okay.

About the Rock House or anything?

Well, I'd love to hear anything you can share about Fern.

Let me put it this way, Gene said. When Fern was in high school our house had a floor furnace and we could hear someone talking when they were downstairs. Needless to say, Fern and her dad didn't get along worth a flip. Fern was a rebellious gal, she was gonna do things her way. She was gonna pick her own friends. She was gonna date whoever she wanted to. It didn't make a difference what he thought. She let him know. It would fire him up. My sister and I had our ears pressed on that furnace listening to every word they said. I don't know what words they used. Mama Faye would be in the other room crying. She went to Oregon and California when Roy was in the War. I don't think Fern wanted to come back. But she did. And raised you, he said to my father. They used to call you Tony the Tiger.

They still do, my father said.

You know, Warren said to me, your dad used to be a basketball star, a singer, popular with all the boys and girls. Growing up, he was my hero.

My father laughed. I didn't say anything, but I remembered that in my boyhood years he'd been like a hero to me, too.

After another round of basketball tales, we got ready to go. Warren wanted to show us around campus real quick. It was late in the afternoon before we wrapped up. On our way out of town, we stopped at a Starbucks. I ordered a coffee. My dad ordered a latte with caramel and whipped cream. I looked at him. He smiled.

As we drove the light faded and we could no longer see the land. We talked about the stories that glossed over what happened and how hard it was to know who was at fault and how the consequences were passed down no matter what. We agreed on the hard truth that Gene felt all the water was worth nothing in the end.

You know what, my father said, I don't know when we'll make it back down here again.

Yeah. I don't know either.

I am glad we came.

Me too. I paused. Thank you for taking me.

In the night the road stretched over the riverbeds. We drove past the Yellow House Canyon and the Palo Duro country and the Adobe Walls and the Buffalo Wallow and other places without names I knew. With two hours left to go we were back on the High Plains. As I gazed at the linear horizons following us I remembered what Gene had said and I pictured the giant earthmovers my forefathers used to level the land for flood irrigation. Later I heard they used to call it knob knocking. The bulldozers planed the prairie flat. They filled in the playa lakes and the buffalo wallows to make a uniform elevation with the correct slope for the water to flow. Sometimes so much dirt was moved that the edges of one field could be eight feet higher than the next. The flatness of the Plains was partly an artifact of vanished waters, although few people can recall the knob knocking these days.

The flat lands lay blank in the darkness, as if we were traveling into a country newly evacuated by some unknown cataclysm or as yet undiscovered. I do not know, then or now, through what space we truly passed that night. And then my father began to sing. He possessed an encyclopedic knowledge of old songs picked up from his grandparents and the radio and his time on the Nashville country music circuit. I never sang with him. But I always liked to listen. That night he started with church hymns and ended with country songs. He repeated the chorus of the last song he sang several times and it said let me watch my children grow to see what they become. And so we moved the last miles north, past the spinning pivots, under the turning stars.

* * *

It was long after midnight when we got back. I lay down in the barn and closed my eyes, dizzy after hours on the road. Fragments of the day's journey and words flashed again in my mind. At the very edge of wakefulness and sleep, Gene's question about marriage returned. As my waking consciousness ebbed away, I was visited by the image of a woman and a child.

In my sleep I once again felt the woman's nearness in my bones. She moved through the dark barn where I lay and when I opened my eyes I could not tell if I was dreaming or awake. Her beauty was incandescent. It shone like a reproach. The features of her face never changed although she resembled an entirely different person from one instant to the next and in all of her unchanging variation somehow gathered a crowd of those I once knew and loved. Beside her hovered a smaller figure, faceless and faint as smoke, the child I would never meet.

For most of my adult life, I could not bear the thought of becoming a father. There was no question: I was not up to the task. I guess I was afraid of doing worse than other fathers, convinced something deep within me would lead me to damage my offspring as I defied any reasonable pressure to conform. I feared putting a child through more pain than I had gone through. Yet the more I tried to avoid acting like the men from my childhood, the more like them I seemed to become.

The news arrived in my dream along with the feelings of an intimacy that had fallen apart. It left me stunned and somnambulant. In my sleep, as moments stretched like weeks, I tried to sort things out on my own. All too slowly, it began to dawn on me, among the confused thoughts of my slumber, that being aware of past mistakes did not mean I was doomed to reproduce them. But it took me too long. I did not find the right words in time and one scene blurred into another as the ghost-like figures began to dissolve. I strained to reach out to them but to my horror I found I could not move or speak and could only watch mutely while they merged again with the dark of the room.

I don't like to think about everything that really happened. When I am awake, I keep my lost chances buried down deep. Some nights they cannot be contained. Hallucinatory visions well up like subterranean waters. Sometimes it is hard to take responsibility. Sometimes it is even harder to shirk it. I live with those I have lost. And I suspect that it is on those bad nights, just before I wake, that I come closest to understanding what running out really means.

* * *

I stepped out of the barn into the cool morning, my thoughts still muddled and nightmarish. Dogs wagged around my legs. Red cattle lazed by green tanks after watering. Songbirds trilled. Irrigation motors droned. The sun hung just above the eastern horizon. I felt its light warmth brush my skin. My vision drifted along the course of the Cimarron. I saw the yellow grass in front of the gray sand hills and the lone elm tree. The tree stood where the riverbed curved back to the west.

As I gazed at the morning scene, the dreams of the night had not entirely dissipated and I suddenly saw filmic imagery of this exact spot, shot in 1957 by some unknown family member. One of the 16 mm film reels, which Gene had given to my father, depicted the same sand hills and river bend from a similar angle. It showed a dozen trees flanking the lone elm. In the foreground of the scene, harvest is underway. The camera follows men now long dead dressed in classic farmer apparel and driving archaic machines across the stubble. RW in a straw hat pulls a metal-wheeled drill behind a roofless Model D tractor. His brother-in-law drives a 1948 McCormick combine with one hand and holds an umbrella with the other. A hired hand in stained denim overalls climbs all the way into a chaff chute until only his booted feet are visible, twitching with the strain.

The images of men and machines are interrupted in the middle of the reel by a two second shot of a young girl. The image is out of focus. She stands facing the camera. It appears as though she is posing. Behind her stretches the vast field. She wears high waisted jeans and a pale blue shirt, with a pink ribbon tied around bobbed dark hair. There is something about the girl's posture and the brevity of the shot that suggests the image was shot at the insistence of the child and not the photographer. On her blurry face is a small smile. Her expression is hopeful and tentative at once. The reel then cuts back to augurs and threshers and worn-out clothes. But it is the afterimage of the smiling girl that hangs over the rest of the scene.

That morning, as I looked at the land, the girl's face crowded my vision too. The footage of old technology and deceased men and that wistful smile floated over the landscape, which now took a form impossible to foresee when the film was shot a half-century before. At that moment, it seemed like I caught a glimpse of hopes from the past and—for just a split second—I could feel in my gut the absence of other futures that were sown in bygone days but that would never grow to fruition. I could not say, then or now, what was actually harvested from the vast fields. I stared at the terrain overrun with ghostly machinery

FIGURE 17

and faded snapshots and the unrealized possibilities that felt ever-more real.

We face a blank future with insufficient means and censored memories. Before I could decide if this was tragedy or mercy or both, the front door creaked open. My father came out of the house. The dogs swarmed. He hollered across the yard get dressed for the meeting, we got to get going.

* * *

The purpose of the board meeting of Southwest GMD was to publicly present and approve a plan to manage depletion. My father and I came in late. It was in the same boardroom where we first met John. We took seats near the wall amid sidelong glances. A photographer snapped shots of John in action for a forthcoming feature article.

I was anticipating a debate. But like other pivotal moments that would decide the future of the aquifer, the gathering seemed strangely flattened and trivial, as if veiled behind some gauzy medium that muffled words and distorted time. The atmosphere was so peculiar that it was not hard to imagine we were somehow already beyond the world. Like my inability to remember the prior meeting, I can recall only a few details from the actual scene, including the facial expression of a board member lounging in a high-backed chair and the highly polished surface of the massive table.

It was only through comparing my notes with the meeting minutes and the public document that I was able to reconstruct, months afterwards, more of what must have transpired in the room. According to those minutes, kept by the secretary, six board members were absent with permission, including the county representative. Nine men from the board were listed as in attendance, along with John, a state representative, a local farmer, and two invited visitors from southeast Colorado. Although my father and I were surely there, our presence does not appear anywhere in the record.

The board members summarized the management plan. My father took a deep breath and hung his hat on his knee. The figures started to wash over us. We both began to fidget. I knew we wouldn't be staying.

John narrated the water problems facing southwest Kansas. He began by repeating that Southwest GMD's mission was, in his words, to conserve and develop the water supply to grow the social, economic, and natural-resource well-being for current and future generations in the public interest. It all comes down to how you define public interest, he said. For the district, public interest was the same as economic growth.

This raises a question, he asserted in the document and his narration of it, about the economic value of groundwater. The text provided an answer. It assigned a dollar amount to water. The State of Kansas loses a total of $3,911 per year for every acre of land that is no longer irrigated, based on yields, taxes, and land values. Therefore, according to the text, switching to dryland farming from irrigated farming is against public interest.[2]

In fact, the text continued, if we use generally accepted economic multipliers, the total value from irrigation was more than $550 million in 2013 alone, and several billion dollars overall. The document then invoked these figures to claim that imposing reductions would be irresponsible. As the text put it, "eliminating aquifer depletion through rapid reductions of pumping limits to safe yield levels could force the transition of a majority of irrigated acres to dry land [and] such rapid change would cause economic collapse that would not be in the public interest." The board was committed to what it called "a situation of controlled decline."[3]

In order to avoid economic collapse, the text reasoned, depletive irrigation must continue. The solution to aquifer loss was not to proactively transition to a more sustainable economic or agricultural system, but to just take water from somewhere else. The document stated the board was developing plans to replenish supply to match a high level of use. Its plan was to replace the aquifer with water piped from the Missouri River all the way across Kansas. In the words of the document, "good, creative, local problem-solving folks have concluded the time is ripe to pursue a major water transfer project."[4] The text argued that available water supplies were inadequate to develop the area's production potential. As evidence, the document cited a study commissioned by the Southwest GMD board. The study, it claimed, found that in the fiftieth future year from 2013, the expected annual loss to the Kansas economy due to insufficient water supply may be $18.3 billion. This, the report mentioned, is exactly the same amount that it would cost to build the aqueduct now.[5]

It was ironic. Aqueducts are a recurring theme in utopian visions of the Plains.[6] As early as 1896, a former immigration agent for the Santa Fe Railroad proposed constructing huge underground reservoirs with gravity-propelled discharge tunnels that would irrigate all the arable lands in the arid West. Such ideas have been resurrected ever since. In 1967, the US Army Corps of Engineers tested a plan to pipe water from the Canadian Rockies to the Plains. That same year, the so-called Beck Plan proposed diverting water from the Missouri River to a canal that would stretch from Nebraska to Texas. Others have tried to import water from the Mississippi, the Ozarks, or the Yukon. Every one of those schemes failed. John was undeterred. He formed an Aqueduct Coalition to work alongside the board and became a national lobbyist for the plan. In private, some folks told me it was a political nonstarter.

The conflict of interest seemed as clear as the consequences. The official plan was to allow the aquifer to be drained and then to replace it with a big-money pipeline. The argument was not so clear in real time. It was only after reading the document several months later that I was able to follow the labyrinthine logics. The more I reflected on them, the more they appeared to border the edge of reality and fantasy. Part of the problem lay in the magic of the numbers themselves.[7]

Like the surreal arithmetic of agribusiness profits, the board's efforts to quantify aquifer loss revealed an illusion at the core of management. Existing models of depletion are not able to numerically contain the sheer complexity of earth systems like an aquifer or problems like depletion that unfold along several dimensions at once. The numbers can tell only part of the story. All models of aquifer decline in the state rely on metadata from WIZARD, the aptly named open-access database operated by the Kansas Geological Survey. WIZARD is one of the most comprehensive data sets on aquifer decline in the United States. Yet even its data are explicitly constrained and incomplete. Measurements of depletion are taken from annual samples of a set of test wells. While these provide important insights at one scale of analysis, there are simply not enough wells or samples or continuous figures to accurately model the aquifer on either a very large or a very small scale. While this is acknowledged by modelers, it is often overlooked or ignored by regional groundwater managers.

This allows for a trick. The patchiness of aquifer systems means that groundwater has unique characteristics in almost every spot. The relationship between any given place and surrounding areas of the aquifer is highly particular and idiosyncratic. Experts at the KGS are aware of this point. Yet Southwest GMD's management document uses this data in ways that erase the layered nature of the water sands. Instead, the document portrays the entire 8,400 square miles of the district as if it were a single, homogeneous unit of water. Despite the immense variation within the High Plains aquifer, the district's arguments for "controlled decline" are based on arithmetic from district-wide numbers for total use, recharge, inflow, evaporation, and storage. These equations ignore subterranean variability, overestimate many local rates of recharge, and artificially disguise the most severe rates of depletion. Treating a patchy set of relations as a single uniform object is a common problem in groundwater management. The risk is that such numbers can pose as a sleight of hand that prevents real conservation in the most threatened portions of aquifer systems. If the aquifer under southwest Kansas is treated as a single unit, eradication is assured for much of the area.

John kept talking. My father nudged my leg. I looked up from the pages. He glanced at the door and back and raised his eyebrows. He sighed but stayed in his seat. As we sat there uneasily, it did not occur to me that the mysterious numbers could be connected to some of the board members themselves. At the time, I took their biographies for granted. It was only a year after the meeting, when I revisited the chart in the Kansas Leadership Council's report that listed the top 150 users of Kansas aquifer waters along with how much groundwater they pumped, that I discovered several of the Southwest GMD's board members or their families figured by name on this list. Others owned, were employed by, or held shares in the agribusiness corporations that were the largest users of aquifer water in the state.[8] A few were extremely wealthy. One board member was a prominent investor in the area's ethanol plants, one's family was a prominent donor to conservative political groups, another was tied to a megadairy, and so on.

Based on this report, I learned that the men gathered around the boardroom that day consumed astonishing amounts of water. The quantities were obscene and the numbers were hard to comprehend. An acre-foot is 325,851 gallons of water. Current law permits each farmer to pump 2 acre-feet each year, per acre of water rights. According to the *Kansas City Journal*'s chart, over the twelve-year period from 2005 to 2017, the operation run by one board member pumped 41,700 acre-feet. That is more than 13,588,000,000 gallons. Another board member's family used more than 56,000 acre-feet. Another, who served on the board for twenty-five years, took more than 57,400 acre-feet. The head of one of the board's advisory committees extracted more than 93,000 acre-feet, and a company associated with him 69,100 more, for a total of more than 162,100 acre-feet. That meant one advisor to the board ran an operation responsible for using more than 52,820,517,100 gallons of water over twelve years. That is roughly equivalent to a column of water covering one acre and stretching thirty-one miles into the sky.

At the time, I was not aware of the close ties between some of the board members and these corporate interests in pumping groundwater. That may help account for my otherwise incomprehensible failure to interrupt or to make even the slightest attempt to perturb the numbing vapors that hovered over the meeting. Looking back, I find it hard to forgive myself for not trying to do more. In my memory, those brief encounters seem like a surreal scene I somehow invented or imagined. As I revisit the images in my mind's eye, I can almost recall the bodies of those present beginning to swell with each muted word and how the fleshy membranes slowly but inexorably filled the room and ballooned against ears and eyes and mouths and bricks until the bloated sacs, by then no longer recognizably human at all, broke through the walls and over the asphalt and past the grain elevators and the meat-packing plants and spread gurgling to the floodplains of the dry river that bordered the town, where they snagged on a patch of dead willows and then ruptured their contents into the absorbent and purifying sands. I am nearly certain that this did not actually happen. Yet I do not know if such events would have been more or less strange than the fact that no trace of the aquifer waters seemed to distort the completely ordinary appearance of these men.

At the end of the meeting, a local farmer objected to the plan. In a halting voice, he asked the board to consider mandatory reductions on pumping. In response, one board member rolled his eyes. Another loudly sighed. A third stated forcefully that talking about reductions was just beating a dead horse before John, with impeccable politeness, buried it all in a flood of phrases seemingly spoken in an alien tongue.

I was growing agitated. My father must have felt the same way. He put his hat on and stood up in the middle of the meeting and walked toward the door without looking back to see if I followed. As we left, they moved to the next agenda item. The words were cut off by the door closing behind us. We stepped from the air-conditioned room into the late spring heat.

* * *

After a stop at the feed store, we filled up and headed back. The road led west along the course of the dry Arkansas River, where the first experiments in western Kansas irrigation occurred. Deflated by the meeting and the bright sunlight and absorbed in our own thoughts, we spoke little.

We crossed the sandy bed near the very spot where, in 1816, a band of fur trappers barricaded themselves on an island behind their bales of beaver pelts and supposedly fended off a force of two hundred Pawnee warriors on a long distance raid. Chouteau's island was gone, disappeared along with this stretch of the famous river and the dropping water table. In the small town that stood along the banks, they had discovered dangerous levels of uranium in the drinking water. We traveled south through ten miles of sand hills and back onto the flatlands that stretched in a level plain to the Cimarron breaks. The land we passed was deep in drought. The weight of the dryness grew more palpable each day, an atmospheric tension that continued to build without ever letting up. At times, it could feel like we were all waiting for something to break under the strain. Mists from the aquifer waters blew under spidery sprinklers.

As we drove, the heat intensified until the edges of land and sky began to shimmer. The heat waves and the pressure reminded me of an unpublished story I had found a short time before in my grandmother's archives. Titled "A Long Wait for Harvest," Fern wrote it under a pseudonym in 1950, when she was in her mid-twenties. It was rejected from several magazines.

The story began with Evelyn lying motionless in bed, staring at the ceiling in her western Kansas house. In Fern's words, Evelyn sensed the oppressive heat closing in, spreading its stifling mantle over the Plains, over the town, and over her troubled marriage. She woke alongside her anger at her husband, an anger that had been there so long it was like a part of her. As she tied a yellow housecoat defiantly around her trim waist, Fern wrote, she noted the tiny frown lines growing between her brows. She made coffee. She did not try to hide her scorn from him and they ate in silence. As she cleaned up, he strode towards a pivotal board meeting that would define his future at the grain elevator, where he found work after the Army. The four gigantic concrete silos of the elevator towered over her tiny house. They were visible from every window. At times they appeared to her like skyscrapers lifting her away from the bleak prairies; at others, they pressed upon her with all of their empty weight. Motherhood felt like the final straw, dulling the luster of her marriage and saddling her with a responsibility that was often more than she could bear. Her husband may or may not have been the father of the child. She watched out of the kitchen window as the board members pulled up in their shiny cars. If only she were the man in the family, she thought bitterly, she would set the board of directors straight. She knew that her husband's demotion to bookkeeper was a trap. The new manager was the son of a corporate grain dealer. He was stealing part of the profits from local farmers. Her husband had meekly decided to go along with the scheme. She spooned cereal into the child's mouth and vowed to herself that she would leave him if he did not show enough strength to stand up to the board members. What kind of an example, she wondered, would that set for the child and generations to come? A man, she wrote, had to fight against corruption and damn the consequences. As she was engrossed in such thoughts, she noticed a black cloud gathering to the northwest. The wind came with howling fury and a splatter of hailstones and savaging rain. She listened to it rage against the tin roof. Her husband didn't return for lunch. Alone in the house with the child, her heart began to race. She wanted to run outside. She felt ill. Tears welled in her eyes. Like a flash of lightning, the realization struck. Wasn't it also her failure to measure up to her own responsibilities that worried her and tortured her as much as the weaknesses she noticed in her husband? She wondered if she had wronged him. She wondered what to do. The storm passed. The air cooled. She found it was the little details rather than the big questions that kept her going. Guilt-ridden, she baked him a Devil's Food cake, his favorite flavor, and met him at the door with a hug. As they stood there, her husband said the board meeting had gone well. The manager underestimated him as a yokel unfamiliar with the ways a big city operator worked. In fact, he kept careful records that detailed the manager's deception. The board members fired the manager and offered him the job. He turned it down. Instead, he proposed moving to the farm. If a man can't put his heart into a job, he told her, then he's only half a man. The farm and the land is where we belong. No place else is as demanding as the wheat country, he said, a faraway look in his eyes. No place takes so much out of a man, she wrote, and no place rewards her own so abundantly. Evelyn laid her head on his shoulder, her wait over.

There, the story stops. Fern's own wait never seemed to end. She kept the manuscript and the rejection slips the rest of her life. She must have gazed at them often. The typed pages were covered with ink drawings and notes and lists of names for characters.

Tucked among her files was a blue binder. It contained hundreds of pages of excerpts from books with titles like *Our*

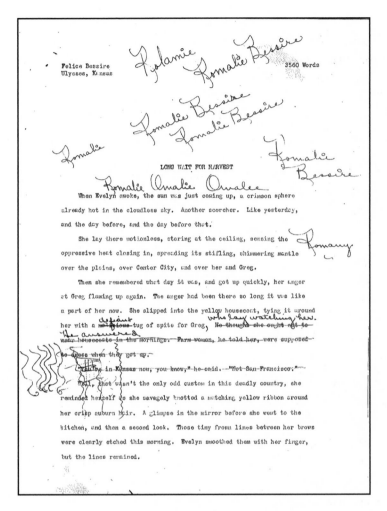

FIGURE 18

Troubled Selves and *Advice from a Failure*. The pages, written
longhand in various colors and scripts, copied inspiring passages and cognitive techniques. Fern broke into her own voice
only a handful times, once in answer to the quiz *What Kind of
Child Were You?*

"I tried to win parental approval by performing well," she wrote. "Yes, I remember raging inwardly, many times. Yes, I was afraid of them. Yes, I guess I still am. I was anxious for their approval. My parents were stern, Dad more than Mom."

> *What made them annoyed with you?*
> Just about everything.
> *What made them laugh?*
> I don't remember them laughing!
> *What did they tell you about your looks?*
> Pretty is as pretty does.
> *What did they warn you about?*
> Being bad. They didn't quarrel at all, ever. They were partial to my brother, of course.
> *What was their attitude if you did something well?*
> It was taken for granted.
> *Did you defy them?*
> Yes, often.
> *How did they express disapproval?*
> PHYSICALLY!

Fern wrote the last word in caps and underlined it twice. Two pages later, in pencil, she posed a question and answered it for herself: "Didn't they punish you enough without you continuing to help? I must overcome the past to realize my goals." Penned in ink above these lines were phrases she most likely copied from a chapter titled "Transcender of Time." The original source of the passage is not mentioned in her notes and I could not ascertain whether the lines were drawn from a published book that truly existed or if she was, in fact, playing the role of the author and quoting lines she imagined herself. "We suffer from momentary lapses in our appraisal of reality," she wrote. "Memories are often less reliable than dreams. We should not give up our fantasies for reality but work to make reality more closely resemble our fantasies," she noted. "One should develop a nostalgia for the future instead of the past. Psychologically and spiritually, man cannot live by clock alone."

Fern was particularly fond of her own clock. The torsion pendulum mechanism was housed in burnished oak. It sat on a shelf above her favorite chair and chimed the hours as they passed. When I stayed with her as a child, its ceaseless motion fascinated me. She got it from a local clockmaker, who kept the delicate gears tuned as part of the deal. His shop was ten miles out of town at a crossroads. My father and I headed there to ask about water. He was on the phone when we walked in. The white walls of the small room were lined with dozens of clocks of all sizes. Most were in old-fashioned styles. Tiny cuckoo sets perched next to the long-case grandfather clocks that were once the height of civilized society. They ticked and moved in an ordered synchrony that was slightly offset. The ticks marked the loss of each moment dozens of times. I wondered how he could stand the pressure. My father nodded to a medium-sized pendulum clock made of carved oak that flared around the top with chimes that hung to the base. That one is just like Fern's, he said. The man hung up and came to greet us. He was tall and handsome and neat, with a well-groomed beard and wire glasses that had an extra magnifying lens tacked onto the right side.

I know you, he said to my father.

This is my son Luke. He wanted to talk to you.

Your grandfather got me started out here, he said to me. He let me take some things on credit. I also knew your grandmother. She was a very smart lady.

Good to meet you, I said. Where did you come from?

Manitoba.

Any chance you have people in Paraguay?

Oh yes. Some of my people came from Russia to Canada to here and then Paraguay. I don't know which part exactly they ended up. He smiled. I suppose they were afraid of the Devil but he kept following them. Now there is nowhere left to go.

I told him I was writing about aquifer depletion and that I wanted to learn Mennonite perspectives on water. Did he know anyone we could talk to?

I have run out of water myself, he said. My well is down to a hundred gallons a minute. I also farmed two quarters on east for another family, but they were down to a hundred gallons a minute, too. One just spat out some water from time to time. We're in a bad water area here but just over there it's good again.

He looked at me. What do you do for a living?

I am an anthropology professor in Oklahoma.

I am not familiar with what an anthropologist does.

I study culture and environments and why people do what they do.

Oh, I am also an expert in culture, he said. There are lots of cultural differences here. People say there are many differences between the United States and Canada. But there are also differences between here and twenty miles from here, ten miles from here, just down the road. Even among our own people there are big differences. Even among us in this shop. Can't culture be anything?

I smiled. Yes, that's right.

Are you a married man?

No.

Are you a religious man?

No. I paused. I guess you could say I am spiritual but not religious.

Don't you think there are certain things that people everywhere know are right and wrong?

I don't know.

Killing and stealing. Isn't that wrong everywhere?

I'm not sure.

Everyone knows killing and stealing are wrong. No matter where you go.

They still do it.

Yes they do. Why do they do it if they know it is wrong?

I have no answer for you.

What is it that makes people go against what is right?

I cannot say. Maybe it is their beliefs or their culture.

There is a line. This I believe. A moral plumb line. It hangs straight down. Like a pendulum on a clock. It separates right from wrong. Good from evil. It is the same everywhere.

Is pumping out the water right or wrong? I asked after a long pause.

Now I have no answer for you.

We laughed.

Well we appreciate it, my father said. We gotta go check on those calves.

The man gave us a name and a telephone number and a firm handshake. We took a back road towards the Little Rock House. In my mind, I still heard the clocks in the shop ticking off moments that were already gone. The road ran by a cluster of rusted structures. I thought it was part of an ordinary silo but as we approached it grew in size until it suddenly took on the appearance of a vandalized temple or defaced pagoda built on land whose ancient fertility had withered into a desert plain. I slowed to look. It was a vast tin-clad ruin. The buildings were spiked with turrets and scaffolds and a tangle of pipes that resembled the exposed roots of a clump of prairie grass. Pieces of tin slapped in the wind and the glass panes were broken. My father said it used to be one of the world's largest carbon black plants.

FIGURE 19

Workers at the plant, owned by a Delaware-based firm called the Columbian Carbon Company, burned natural gas under large metal frames. They scraped off the soot and sold it to other companies for use in pigment or rubber tires. They called it cracking the gas.

In the late 1950s, farmers near the plant reported a veil of fine black grime hanging over the land, disrupting their operations. One farmer said this mist of black covered all the plants, buildings, fences, crops, and livestock with a rank and foul-smelling film. It turned rain into rivulets of ink. It soaked into the soil. It stained their wheat. Farmers called it "living in the black."[9] They filed a dozen lawsuits for damages. The company denied all charges. Farther down the road was a small rural community where, in the late 1970s, they noticed that 30 percent of the children were born with severe birth defects. The families of the afflicted children lived within two blocks of one another. Their only common tie was the mysterious fog that, according to them, drifted over their houses after nearby fields were sprayed.[10] The mayor requested a state inquiry. Aerial sprayers at first confirmed that numerous chemicals were used in the area. Among the applied substances was 2,4,5-T, a defoliant in Agent Orange that also contained tetrachlorodibenzodioxin.

This chemical, linked to miscarriages and birth defects else-where, is a potent carcinogen in animals and humans. It lingers in the soil and water for decades. In response, the Environmental Protection Agency asked dealers and farmers to stop using 2,4,5-T on crops. Many farmers objected to the suspension. The sprayers walked back their admission. The document trail stopped abruptly in 1980. I supposed it meant a settlement had been reached but I could not find out how either the lawsuits or the contaminants were resolved.

Such stories hover over the land like dust. They cast doubt on the consequences of the chemicals that Plains agriculture relies on. If most farmers acknowledge the dangers of running out of water, there is no such consensus over the ways agrochemicals further deplete our bodies and ecosystems. The risks they pose to health and to future generations is often a taboo topic among residents. And it seems as though there is an intentional effort to cast doubt on the degree to which these toxins cause bodily harm. There are rumors of corporate attacks on independent research about pesticides and smear campaigns targeted at critical voices. Paid experts and lobbyists for powerful industry groups create uncertainty about chemical causation. They portray lower thresholds of exposure or greater regulation of certain chemicals as a threat to rural economies and liken it to a form of persecution against local communities. If the implications of these chemicals are a matter of debate, few can deny the extent of their use.

The soil of the Plains is saturated with chemicals. The constant wind often blows this dust and the spray far from its targets. No-till farming—often promoted as a way to slow evaporation and to conserve groundwater—takes chemical application further. It means chemically killing all plant life then implanting new seeds below the dead surface matter. Atrazine is a critical ingredient for no-till farming. It is also applied to 60 percent of the nation's corn crop and 75 percent of sorghum. The problem is that it can also make its way into water supplies. According to some research, the compound is an endocrine disruptor linked to birth defects.[11] Other agrochemicals are also suspect. Some scientists have correlated exposure to pesticides—including several kinds of organochlorines and organophosphates—with some chronic illnesses that are widespread among residents in western Kansas.[12] These include obesity and diabetes, as well as pulmonary disorders and non-Hodgkin's lymphomas. Some agrochemicals appear to interrupt basic cellular functions, interfere with hormonal paths, or change the bacterial flora in the human gut. Some researchers have also tied pesticide exposure to neurological damage. It is possible that some of the chemicals that drench the Plains may further deplete our already shaky capacity for reliable memories.

* * *

A few miles south of the ruined factory the road dipped into the Cimarron Valley. The hillside was terraced in the 1930s. The faint scar of an older excavation cut across the bottomlands. It was all that remained of a failed attempt in the late nineteenth century to dam the Cimarron and drain its waters for irrigation. The South Fork Irrigation Corporation formed in 1894. That year, several settlers dug a ditch across the valley and aimed to divert forty cubic feet of water per second from the river's flow. At the time, nearly every stream was associated with one or more canal companies. Many represented out-of-state corporate interests. Canals like the Southside and the Amazon were dug on the Arkansas River, some sixty miles north. Newspapers

FIGURE 20

declared such canals would "redeem the desert" and turn southwest Kansas into an Eden.[13] Ditches were also planned for
smaller streams like Bear Creek and, just across the Colorado
line, Sand Creek. In 1908, the Chivington Canal Company
drained Sand Creek's waters for cultivation.

Once, Fern took me to visit Sand Creek. I recall a long trip
on dirt roads, south of the town named Chivington and not far
from the Camp Amache site where several thousand Japanese
Americans were interned from 1942 to 1945. At the time of our
visit, Sand Creek was on private property. There was no sign or
marker to indicate the spot. Fern must have called ahead. The
day was hot and windy. We walked across the pasture to a fence
post, where a wooden box hung with a handwritten sign re-
questing a two-dollar donation. A short distance from it stood
a pink granite marker. The stone looked like a tombstone and
was exactly like the one at Wagonbed Springs. It was engraved

with the image of a man in a headdress and captioned Sand Creek Battle Ground. It stood on top of a small bluff. The dry creek curved around a flat meadow. I distinctly recall the uneasy feeling of the place. We stood in silence. After a short time, we walked back to the truck and headed home. I do not recall what Fern must have told me about the events that occurred there. I am certain she did not explain the details. But she undoubtedly knew the broad outlines of the story.

In November 1864, a peaceable group of southern Cheyenne and Arapaho camped in the bend of Big Sandy Creek. They were led by Black Kettle and other clan chiefs. Several months before, some of these same leaders had brokered an exchange of captives and saved the lives of soldiers under the command of Ned Wynkoop. The chiefs traveled to Denver with Wynkoop, where they met the governor and military authorities, including John Chivington, a Methodist preacher and abolitionist who commanded the Colorado Volunteer militias. The chiefs asked where they could live in peace. The governor promised them protection and provisions if they stayed near Wynkoop's post at Fort Lyon, just over the Kansas line in what is now eastern Colorado. Wynkoop gave them his word.[14]

Genocidal sentiment ran strong on the Plains. Chivington complained to his commanding officer that Wynkoop was too sympathetic to the Indians. Wynkoop was reassigned. On November 26, Wynkoop left Fort Lyon. On November 28, Chivington arrived unannounced at the fort with around eight hundred soldiers, most of them volunteers signed on for hundred-day contracts to the newly formed Third Colorado Cavalry Regiment.[15] When Chivington arrived, he put a barricade around the fort and threatened to shoot the officers who were loyal to Wynkoop. The new commander told the officers they were going to fight hostiles. After a few hours of rest, the men rode through the night, directly to the friendly camp. They brought four mountain howitzers that fired explosive shells weighing twelve pounds each. They arrived near the camp before dawn. Chivington ordered them to strip for battle. He gave them a short speech. He said they should picture the blood of White women saturating the sands of the Platte.[16]

The camp consisted of around 110 lodges and perhaps five hundred people. Most of the men were away hunting buffalo. Two-thirds of the inhabitants were women and children. The soldiers attacked at sunrise, "a perfect mob." A Cheyenne woman heard them coming. Black Kettle hoisted an American flag. One White trader ran out with a white flag on a stick and was almost shot. White Antelope walked towards the soldiers yelling Stop at them in English. Others kneeled and held up their hands. They were shot or beaten to death. The soldiers opened up on the village with the howitzers. "People broke and ran all over the ground."[17] The main group of Cheyenne ran up the creek about two miles, soldiers following them the whole way. Eventually, they dug in under some steep banks. At nightfall the soldiers returned to the camp. George Bent, shot through the hip, was one of the survivors. He recalled the way they crawled up the creek and camped on the open plains. "That night," he wrote, "will never be forgotten as long as any of us who went through it are alive."[18]

Roughly a quarter of the people were killed. The soldiers collected souvenirs after the carnage. They mutilated the dead. They burned the village. They executed one of the captives. They rode back to the fort festooned with scalps and genital organs, which they hung on sticks and draped across their hats. They took the scalps back to Denver and displayed them in the theatre and made tobacco pouches from the skins. The Denver newspapers celebrated the massacre. One editorial in the *Rocky*

Mountain News called it a "brilliant feat of arms [in which] Colorado soldiers have again covered themselves in glory."[19] Chivington was told he would be promoted to brigadier general.

Several officers protested before, during, and after. Captain Silas Soule, one of Wynkoop's men, called them all "cowardly sons of bitches." During the attack, he ordered his company not to fire on the camp and he moved them away from the killing ground. Afterwards the volunteers threatened to shoot them or any man who "shot off their mouth about killing Indians."[20] Two weeks later, Soule wrote a letter to Wynkoop about the horrors. "You would think it impossible for white men to butcher and mutilate human beings as they did there," wrote Soule, "but every word I have told you is the truth, which they do not deny."[21]

Wynkoop and Soule filed complaints at the War Department in Washington. Due to three federal investigations that were launched in February, details of the slaughter entered the historical record. Soule and others testified. Chivington was allowed to cross-examine them all. He accused them of trying to "stain the glorious field of Sand Creek." The Denver papers denounced the investigations as "bosh," a "conspiracy of effete Easterners" who deserved to be shot and scalped for protecting "marauding bands of savages."[22] Soule knew he risked his life. Two months later, he was murdered in Denver. The congressional commission harshly criticized Chivington. But he was never charged with a crime and remained unrepentant the rest of his life.[23]

The massacre unleashed war across the Plains. The slaughter was repeated in 1868 along the Washita River. After 1870, most southern Cheyenne were confined to Indian Territory. In 1875, they were joined by the Comanche and Kiowa survivors of the

Buffalo War. For decades afterwards, White soldiers and ranchers traveled back to Sand Creek to pick up relics from what they called the "desolate looking place."[24] Some bones were sent to the Smithsonian as evidence for what bullets did to skulls.[25] Many of the eye witnesses moved on or died off. In 1908, four veterans of the Third Regiment visited the area, guided by a local cowboy and accompanied by a reporter. Each man thought the massacre had happened in a different place. The reporter concluded that nobody was sure where it actually occurred.[26]

Waves of settlers occupied the land. Many knew little about the area's recent past. In 1950, the granite marker seemed to dispel uncertainty about the location of the massacre. Amateur artifact hunters and archeologists, however, failed to find many traces nearby. By 1998, National Park Service officials testified that the exact location of Sand Creek massacre was unknown. Bitter debates between local ranchers, tribal authorities, Park Service employees, and archeologists resolved little. Some believed the massacre occurred below the monument. Others were convinced it happened farther up the creek bed, where archeologists did find artifacts. As historian Ari Kelman chronicles, the massacre assumed a ghostly presence that floated over the land.[27] It was nowhere and everywhere at once.

Perhaps that is somehow related to the ways Plains children like me were able to indulge our curiosity about the past without ever connecting it to brutal atrocities or to present attitudes about agriculture. We lived among the rubble of genocide and dispossession in a landscape that had been transformed. Nothing seemed as fascinating as the chips of flint and arrowheads and old bullets and potsherds secreted among the last ribbons of native shortgrass. And nothing seemed as innocent. Only now do I see their allure was part of displacing the monstrous events that allowed me to inhabit the Plains.[28] We confined the

FIGURE 21

horrors of eradication to a cartoonish lost world; one that we thought was entirely disconnected from our own. We did not relate past events to the banal activities of irrigation farming or the way we grew up or the pumping of the subterranean aquifers. Like the extermination of buffalo and the toxic fogs and the torturous confinement of defiant voices, these events were not openly discussed and their remnants were never tied to the present. Cordoning them off from conversation meant that their significance was largely blocked from our memories, too.

The Little Rock House is a two-day walk from Sand Creek. Some of the survivors fled to the Cimarron. There was a continuous Native presence along the river and its springs. Other massacres yet unnamed occurred there, of Cheyenne, Kiowa, Comanche, and others.[29] Fern was certainly more aware of these events than many local residents. She collected files of what she called Indian stories. She wrote letters to local residents and catalogued every artifact found in the county. She took me to reenactments of battles and treaties. Yet she never addressed the violence directly as far as I know. One of the only references she made was on her radio program. No, she said, there were no Indians here when the country was settled by White men in the 1880s.

I puzzle over this statement, which she undoubtedly knew was not true. Was she referring to what was left in the wake of Native dispossession by settlers or was she a party to the tacit agreement that the true state of destruction was not to be described, or both at once? It seems the most gruesome aspects of the processes that cleared the Plains for agribusiness remained under a kind of taboo. Segregated from decent everyday life, they were permitted to linger within it and rise to the surface over and over again. Like a shameful family secret. I am left to wonder why she took me to Sand Creek at all.

According to George Bent's letters, traditional Cheyenne believed that Plains waters were sustained by Mehne spirits. The horned serpents lived in the waters they guarded and were dangerous to humans unless they were appeased with offerings. Without the spirits, the waters dried up.[30] These beings were agents of the feminine power that pervaded the Deep Earth. Some observant Cheyenne also called this power Grandmother.[31]

My father visited Sand Creek in the mid-1990s, on his way back from a singing gig in Lamar. He walked past the monument and down to the creek bed. There, he picked up a round stone, put it in the back of his truck, and promptly forgot about it. Bad luck of all kinds ensued. He started having nightmares. He knew something was wrong but he did not associate it with the round stone until he saw a story in the local paper about the anniversary of Sand Creek, which featured a photograph of the very same stone he had taken. The next weekend, he made the two-hour drive back. He saw no animals or birds. He walked to the dry creek. There, he replaced the stone and added a handful of flint chips from the Cimarron along with some chewing tobacco. Leave me alone, he said out loud. As he walked back to the truck, he noticed hawks spiraling above. His bad luck ended. Later, he wrote a song about it.

I was less sure that such things could ever be put back. I did not know the right place and I never learned any absolving words. The Sand Creek Massacre National Historic Site was inaugurated in 2007. By then, the controversy over the massacre site had taken a new twist. Historians working on the case rediscovered the Chivington Canal Company and its ditch.[32] Around the turn of the century, the company built a series of dams in Sand Creek. Just below the former camp, they dug a canal east from the stream to a reservoir. Promoters said the canal would provide water for twenty thousand acres of alfalfa, melons, and sugar beets. Water would transform the evacuated Plains into productive economic ground.

The Chivington Canal operated for only a handful of years. But the failed irrigation scheme was enough to alter the land. The ditch created a new bend in the stream and rerouted the course of Sand Creek.[33] It also displaced the violence and occluded memories of it. In that place, at least, the deep ties between perception, genocide, and agribusiness were starkly laid bare. The more I thought about it, the less certain I became that what the motors pumped to the surface was ever water alone. And who can say what is truly left behind when a pocket of groundwater is pumped dry?

* * *

After a year in search of water, I knew running out had deeper roots than I could ever have imagined. It was too late to turn back. The layers of aquifer loss revealed uncomfortable and perhaps insoluble truths about the destructive habits at the core of my existence. They demanded a reckoning and blocked the conditions for one at the same time. I did not know how to proceed. It was tempting to give in to despair or to give up and move on. Every time I got lost in the mazes of depletion, shades of Fern seemed to reappear and show me some way ahead, however tentative.

The last part of Fern's blue binder was filled with passages she copied from behavioral therapy texts. She focused on techniques aimed at reversing neuroses. Her notes contain long lists of mistaken perceptions, alternate versions of them, and corrective behaviors. I cannot imagine her struggles but it is clear from her selections that she was trying to realize another version of herself and a more inhabitable version of the world. On the bottom of the final page, scrawled in pencil, was a note she wrote to herself. "No more artificial, useless and degrading escapes. No more solitary drinking. Then booze and sleeping pills. Then cigarettes. The most destructive habits must be the first to go. Remember by giving in to easy temptation I don't change things a bit or make any of it cease to exist." A gap, then she continued in blue ink with a shaky hand. "The first step in my own reclamation of command," she noted, "is to admit that I am not responsible for the past but that I am accountable to tomorrow."

I picture her sitting alone at night, copying cliché after cliché as the sound of the irrigation motors drifted on the wind and the waters ebbed. The techniques never really worked for her. They were not up to the task and I suppose neither was she. Her notebook showed me that she never stopped trying. She refused to stay within the lines others set around her and she did not conform. It may not seem like much. But the more time

I spent on the Plains, the more I came to appreciate her strength and courage. She spent her life spinning between suffering and redemption, without an ultimate resolution and without ever giving up. The search for water became her compass.

After the hospital, Fern doubled her efforts to relocate the site of the Lower Springs. Her files show that in 1937 the local 4-H club placed a marker for Wagonbed Springs south of the riverbed. This area was named a National Historic Landmark in 1960. Fern doubted it was the correct spot. To challenge the federally designated site, she formed a local chapter of the Santa Fe Trail Association. She served as its president. Her main achievement was to create a controversy over the location of the Springs. It was only in the course of writing this that I learned her efforts did, in fact, pay off. The controversy she started led the government to reassess the location of the marker. Seven years after her death, archeological surveys confirmed her local narrative of the past. Based on these findings, in 1998 the National Park Service revised the boundary of the Lower Springs National Historic Landmark to include the relocated Springs.[34] Although the waters flowed only as memories, Fern had somehow found the life-giving Springs after all.

CLOUDS

Some farmers are trying to use present tools to slow the loss of the aquifer waters. Shortly after the board meeting, we followed up with a farmer named Dwane. The former army specialist wore work clothes and spoke with an irreverent modesty. From his clothes to his talk, he seemed as real as it gets.

At the time, Dwane farmed several thousand acres of his own ground along the Arkansas River, some fifty miles north of the Little Rock House. He managed much more as a tenant, including one of the so-called Water Technology Farms. Subsidized by the state Water Office, supported by Kansas State University (KSU) extension, and partly sponsored by technology companies, the farms showcased the most efficient irrigation systems.

The Water Tech Farm began in 2015 when a landowner challenged him to do more with water technology. By early 2016, Dwane had forged partnerships with KSU and technology companies. Together, they installed three of the most efficient delivery methods—mobile drip lines, bubbler nozzles, and wobbler nozzles—along a single pivot. Each applied water at low pressure, which reduced wheel tracks and evaporation. Dwane spaced them at thirty- and sixty-inch widths along the spans. That meant he had six different configurations working on the same field at the same time. To monitor the results, he and his partners put meters in the pivot and telemetric soil moisture sensors in the ground.

The readings were synthesized in a computer program accessible from his smartphone. Dwane tracked root growth and moisture profile in real time. KSU extension officers helped compile and analyze the feedback. A mobile hub allowed him to turn the sprinkler on and off remotely. The input let him know when and how he needed to water; the technology let him put the data into action. It also gave him the confidence to shut down his sprinkler and water less.

That year, Dwane recounted, everybody else started watering. So I thought, start my pivot up. But the soil moisture probe guy called me up. Based on his readings, he said you don't need to water. First I tried to argue with him. But then I took the leap of faith. I shut it off.

The results were eye opening. In 2016, a relatively wet year, Dwane applied only six inches of water and grew more corn than the year before. The following season was even more shocking. He applied less than half of the amount of water used by some of his neighbors. But his crop outproduced theirs by around ten bushels per acre.

Farmers know that many variables shape yields. The major difference between Dwane and his neighbor was the amount of water. He asked around. Too much water, he heard, could push nutrients away from the roots. The sensors also showed that prewatering was not necessary. Even during one of the driest winters on record, Dwane's soil had enough moisture to get by without advance watering. He could do more with less. This led him to an uncomfortable conclusion. He was actually overwatering his crops. Not by a little, he told me. By a lot.[1]

Technological advances in irrigation, of course, change possible relationships with the remnants of the aquifer. Innovative new technology offers important ways to save water and to apply it more efficiently. But the irony is that more-efficient

FIGURE 22

technology can also be used to further depletion.[2] Much of the newest irrigation technology is specifically adapted to continue drawing from depleted or low-output wells. Ten years ago, a farmer would have had to stop pumping from a small well. Now, the same farmer can use more-efficient systems to keep drawing water out of a depleted area. And newer technologies can also access the deeper layers of water sands that were previously inaccessible. Recognizing this, Dwane decided that the adoption of more-efficient technology needed to be coupled with collective action and reductions.

The turning point came when his daughter, then a freshman in high school, became concerned about the future of the aquifer. She decided to do her Future Farmers of America (FFA) project on the topic. She inspired him to commit to the future. They began to work together. "My dad is all about the next generation and he wants to help the future," she said in a local

newspaper article. "I'm part of that future."[3] They said the aquifer brought them closer.

In December 2016, the father and daughter team met with
experts from KGS to discuss what could be done. Dwane's
daughter began speaking to youth from across the state. She
organized a Water Boot Camp for high-school FFA students. At
the same time, Dwane began organizing local farmers. He
hosted meetings and public information sessions. He set up
regular conference calls with experts from KGS, KSU, and the
Division of Water Resources. To his surprise, nearly everyone
he talked to shared his concern about the aquifer.

By early 2017, Dwane had created a group of a dozen local
farmers. They decided to focus on a local enhanced management area, or LEMA. Kansas water law has three mechanisms
to restrict use in areas of decline. WCAs are the least intrusive.
They are nonbinding voluntary agreements between a water
user and the state to limit use and get more flexibility. Intensive
groundwater unit control areas, or IGUCAs, are the most intrusive. They provide a mechanism for the state to impose and
enforce nonvoluntary restrictions. LEMAs fall somewhere in
between. Like WCAs, they are local initiatives to voluntarily
limit use. But like an IGUCA, their limits are enforceable and
binding on all within the area.

The process for creating a LEMA is supposed to be straight-
forward. Local farmers propose it to the GMD. The GMD de-
velops a plan, holds information sessions, and submits it to the
state Division of Water Resources (DWR). The DWR reviews
the plan for legal compliance, holds at least two public hearings,
and then decides to implement the LEMA or not.

Dwane's group proposed a LEMA for a portion of Southwest
GMD along the Arkansas River valley in Kearney and Finney
counties. The proposal became known as the Kearney-Finney

LEMA, or KFL. That area was particularly hard hit by water loss. It was the site of the earliest irrigation ditches in western Kansas. By the mid-twentieth century, stretches of the river no longer flowed continuously. When the river did flow, the surface water was often saline or high in minerals. As the river declined, groundwater pumping grew. Along with the area around the Little Rock House, this portion of the district had some of the highest measured declines in aquifer levels from 2012 to 2017.

Dwane's group started calling itself the KFL Steering Committee. Its goal was irrigation sustainability. The KFL Steering Committee met every two weeks throughout spring 2017. The group solicited feedback from area farmers in meetings and mailings. It debated different restrictions and mechanisms. It adjusted based on dialogue with farmers. Finally, the group shared the plan with the GMD board, who had to approve it before it could be sent to the state. The board did not approve the plan.

By November 2017, the Southwest GMD had decided a survey was required to assess local opinion. The future of the LEMA depended on the results. They mailed 412 surveys to water users in the LEMA area. They got 72 back. Responses were reported to be evenly split. Roughly a third said they were in favor, a third said maybe under certain conditions, and a third said they were against. The GMD board decided it would not act until someone presented a plan with more support.

The story challenges the often-repeated idea that conservative farmers are uniformly opposed to regulation or restrictions on use. In fact, I discovered that the opposite was often true. While some farmers were committed to eradicating the aquifer, far more were not. Most farmers I met in southwest Kansas said they were in favor of fair limits that would address the slow-moving crisis. Moreover, they all expected that authorities would act and were surprised the state had not already taken stronger actions. My impressions were later confirmed by an eye-opening survey of producers carried out by sociologists Stephen Lauer and Matthew Sanderson. In 2018, they surveyed a representative sample of 1,226 farmers from 227 counties across the six states of the Ogallala aquifer region. Nearly all— 92 percent—agreed that the aquifer waters should be saved from depletion. More than half said the well-being of future generations was the reason to conserve. At the same time, most farmers acknowledged that they felt individual efforts could go only so far.[4] Many expected—and hoped—that the government would take effective action to prevent a catastrophe.

What, then, is preventing more effective conservation of the remaining groundwater in southwest Kansas? One part of the problem seems to lie at the midlevel bureaucracy of water management and the implementation of policies that cannot resolve the core problem of an economic system based on extracting a nonrenewable resource that is running out.[5] Southwest GMD's policy of controlled decline can continue only if a majority of voting members permit or support it. Presenting their constituents as a unified and grassroots front is also important for political optics at the state level. In other words, it is in the interest of a prodepletion GMD board to mute or to discredit the voices of proconservation farmers, to amplify the voices of those in favor of eradication, and to make the opinion of one local faction appear as a district-wide, democratic consensus.

Continued decline is not the only possible policy choice for regional groundwater districts in Kansas. In fact, Northwest GMD, which covers the northwest corner of the state and uses a fraction of the water consumed in the Southwest GMD, has developed a policy that seems diametrically opposed to that of Southwest GMD. To do so, Northwest GMD used the same mechanism proposed by the KFL Steering Committee and the GMD's successful efforts happened at nearly the same time. In February 2017, two proconservation farmers were elected to the Northwest GMD board. Afterwards, the GMD worked with local farmers to develop a LEMA, in order to reduce decline rates and extend the life of that area's aquifer.[6] While the KFL Steering Committee sought to restrict extraction in one particular part of the Southwest GMD, the Northwest LEMA covered the entire district. To set it up, the board divided the district into zones based on the rate of decline. They imposed different restrictions on each zone. After factoring in net irrigation requirements for crops and average past usage, they assigned each farmer a five-year total allocation.[7] From start to finish, the process took ten months. A handful of residents filed two lawsuits that challenged the legality of the LEMA mechanism in general. One was dismissed by the court. The LEMA stood.[8] While the imposed restrictions might not result in zero-loss agriculture, they were an important start.[9] The unavoidable conclusion is that the different results were not due to fundamental differences in opinion among farmers, but instead to the opposed attitudes towards depletion held by the two neighboring GMD boards. The difference in these attitudes may correlate with corporate ties and influence. Regardless, advocating for controlled decline makes Southwest GMD an outlier in the wider Ogallala region. And this policy might prove unpopular if its actual workings and implications were more widely known.

* * *

To learn more, we decided to visit one county representative on a GMD board. We knocked on the door of a well-appointed farmhouse. A handsome man greeted us politely. On our way into the house, he said softly there was no need for introductions. He had already heard about us and about the book I was writing. He knew where we stood.

We sat around the shiny table as the representative, in a low calm voice, laid out reasons for optimism. Southwest Kansas farmers, he told us, are the best people in the world and we are already on the road to sustainability. There was no need for restrictions. He was opposed to reductions and to the state's definition of conservation. He was in favor of what he called real conservation based on actual situations. He told us not to believe what we heard and suggested that some proconservation farmers, who he listed by name, were out for their own selfish gain at the expense of their neighbors.

According to the representative, the state needed to reward people for the real conservation they were already doing on their own. Regulating water would just damage the economy and hurt young farmers. He invoked a stream of numbers and figures about property values and tax revenue and decline rates. He did not mention they referred to different scales and could be interpreted as evidence against his argument.

You know there are other numbers that say the exact opposite, I ventured.

I know, he replied, a slight smile creasing his even features. All I do is talk about water. I have a degree. And I have practical experience, he said, invoking the common Plains rebuttal that real world experience counts more than formal expertise.

I know about conservation from watching the creek over there, he continued. We had a two-inch rain and it all pooled up in my terraces. It used to run into the creek, but now it gets sucked up by the CRP grass.[10] Irrigation is not to blame, he said.

I could not tell if it was meant to be a joke. But the representative seemed serious. He urged us not to listen to any naysayers and continued with a series of punchlines. We are blessed with good saturated thickness. There are hundreds of years of water left. There are no de-watering incidents really. In fact, he concluded with a grin, we may just be too good at conservation.

Well, I said. Come on now.

My father looked over and stirred before I got myself into trouble. Thanks for your time, he said. We gotta get back and check on those calves.

As we drove away from the farmstead, I tried to comprehend what I saw as shortsightedness and bland falsehoods that aimed to break perceptions of truth and causality. If that was the attitude of those charged with stewardship of the life-giving waters, the forecast was bleak. I watched the heat waves dance. Along the very edge of the horizon appeared an oval of blue. It looked like a pool of water or distant lake. Plains travelers have long remarked on this optical illusion, where heat and light take the form of ever-retreating water. Such mirages were commonly reported near the same route we drove that day. Crossing this very spot in 1831, Josiah Gregg noted in his diary the optical deception he called the false ponds.

"Even the most experienced traveler is often deceived by these upon the arid plains," Gregg wrote, "where a disappointment is most severely felt. The thirsty wayfarer espies a pond—yes, it must be water—it looks too natural for him to be mistaken. He quickens his pace . . . but lo, as he approaches it recedes or entirely disappears; and when upon its apparent site, he is ready to doubt his own vision—he finds but a parched plain under his feet."[11]

Fifty-six years later, an unnamed buffalo hunter traveled the same area in pursuit of the last remnant band of bison on the upper Cimarron. "In a hot dry country," the unnamed hunter wrote, "it is distressing to see in the distance a lake of water with

FIGURE 23

a fairy sailboat riding on it, reflected in the sky. Whole forests appeared of fine green trees and small villages were seen suspended in the clouds, rivers could be traced and I have witnessed a complete farm scene, shown as if by magic."[12]

Coming back from the representative's house, I stared at the mirage that hovered in the sky like an inverted reflection from the surface of rivers that no longer existed and could not be certain if what I saw was the mirage or the memory of a prior reflection or if I was visualizing a natural manifestation of the representative's words. The extreme politicization of perception has made it nearly impossible to locate shared criteria of truth, evidence, and sense. Confining facticity to opposed partisan worlds renders rational argument ineffective and narrows shared concern. Powerful elites twist the language of decency in order to steal from the future with no plan for repayment other than the debt itself. They bend the words of justice to unjust ends and invoke truth to spread lies. People are primed to reject different viewpoints and stand ready to violently defend imperiled homelands, even as social isolation, economic strain, and droughts drive a mental health crisis in rural areas around the world.[13] Depletive industry flourishes along these fault lines that turn us against the future, against each other, and against ourselves. What does it mean, then, if depletion itself may soon be one of the few common experiences that are shared by people on the Plains?

* * *

Such thoughts occupied me all the way back to the Little Rock House. When we got there, my father said he wanted to work the red cattle soon. We walked out to the corrals to take a look. He had a new bunch of calves. They had to be separated, cut, and ear tagged. He should pull some tail hair off the bull for a genetic test. The cows were due for shots. By then, my father didn't have to ask if I would help. He knew I would be there, even though I wasn't much of a cowboy and would probably just get in the way.

I need a new chute, he said. This one's no good.

We stood and looked at it. The pipes were rusty and the headgate rollers stuck and half the brackets on the foot panels were broken. The alley itself was cobbled out of ancient boards and sheep wire. It didn't look like it could hold a goat.

Yep, I said.

You know I asked Travis if he knew somebody who could build me one. He put me in touch with a guy named Juan. He came out here to talk about it.

Will you get him to build you one?

No. I don't need one that good. But you'll like this. He told me he knew someone who used to live out here.

Oh yeah?

Yeah. Juan said he hadn't been here but he recognized the place from this other guy's stories.

Huh.

Said he'd heard about this house down in the river valley past two hills out in the middle of nowhere and when he drove out he knew this must be it. Said the guy loved the place and always remembered it.

Who was the guy?

I don't know. I can call Juan and find out. You wanna go talk to him?

Well yeah. I thought RW had always owned the place.

His hired hands lived out here too. So did Fern and me and whoever. Two Mexican families worked for him for years.

Who was that?

The names I know are Rodriguez and Ortega.

A couple of days later Juan called back. He had gone over and talked to the family and a woman named Susana gave Juan permission to share her number with my father. He called her up and said I was writing a book about family history. She started to cry on the phone. She said she was thankful to RW for giving her a place to live and that she spent some of her best years in the Little Rock House and that we were welcome to come by. It was the first time I heard anyone say anything nice about RW.

Her red-roofed trailer sat at the edge of town and faced an irrigated wheat field. We walked up the wooden steps and knocked. It was opened by a short woman with white hair and sharp eyes. She hugged us both and led us to the kitchen table. We handed over a pecan pie. She poured coffee in porcelain cups and cut us large slices on plates. Then she sat down and folded her hands and started telling stories about our intertwined families that I had never heard.

FIGURE 24

Susana acted as if we knew as many details about her family as she did ours. She graciously filled in the details. They had lived at the Little Rock House from 1970 to 1978 when her husband worked for RW. Now she was raising three grandchildren, the youngest was eight. As she was talking, her son joined us. He wore an orange hoodie under a Carhart coat. He said he remembered the Little Rock House. He recalled the cowboys who would visit on their days off to play guitar and sing and drink and jackpot rope in the old corrals. He described the rattlesnakes and the ghosts that roamed at night and the time his father walked out in a blizzard to save him and some other kids stuck in a tractor and how his father would crack them up by imitating all the dumb things RW and Gene said to each other over the CB radio.

I sat at their table, sipping the good coffee and listening to laughter, and thought about the strange exclusions inscribed

into my family. Southwest Kansas has always been deeply multiethnic and multiracial. It was once part of the Spanish empire
and Mexico. From Susana and her son, we learned that some
Mexican American families had lived in the area for a century.
Their descendants had brokered contracts with seasonal migrant farm laborers, who lived in temporary shacks set up on
the edges of fields and labored by hand or hoe. Few records
were kept of their numbers or names or destinations. Another
group of Mexican Americans, like the Ortega family, came for
agricultural jobs in the 1950s and stayed for generations.[14] More
recent immigrants came from Mexico and Central America for
work in the slaughterhouses and hog barns.[15] In the last census,
nearly half of the county's population was listed as Hispanic.
Undocumented folks meant the real number was higher still.

Half of my schoolmates were Mexican American. I grew up
eating tamarind sweets and lime salt. I went to *quinceñeras* and
charro roping and learned *ranchero* dance steps. My mother had
lived in Argentina as a child and taught us a handful of Spanish
words. But few non-Hispanic locals spoke any Spanish at all
beyond slang and obscenities. When I was growing up, I took
those racial and class boundaries for granted.[16]

Irrigation farmers are the privileged of the Plains. Nearly all
White, most local irrigators are descendants of the settler, Mennonite, and Volga German families that stayed through the Dust
Bowl. There are very few Hispanic landowners. Most are laborers. As such they are exposed to greater harms from toxic agriculture and dangerous jobs in the meat-packing plants.[17] While
the industrial centers of southwest Kansas are distinctly Hispanic, the area's working classes come from a very wide range
of ethnic, racial, cultural, and economic backgrounds. Southwest Kansas has long been a resettlement center for refugees
from many areas of Asia and Africa. These displaced peoples

join members of several Latin American Indigenous groups, *Plautdietsch*-speaking Mennonites, European immigrants, and Exoduster-descendant Black communities to occupy the lowest paid agro-industrial jobs, primarily in the big meat-packing plants in Liberal, Garden City, and Dodge City. There were fifty-two languages spoken by students in my high school. Poverty rates roughly reflect the general demographics, with so-called Whites accounting for around 60 percent of the area's poor. These working classes will face harsher consequences of aquifer decline.

Because only landowners can vote on groundwater policy, nearly all working people, Hispanic or otherwise, are excluded from the process of making decisions about the aquifer. It means the current structure of water governance is antidemocratic. And it also means there could be an opening for change. If residents and townspeople were included in groundwater management, it would alter the decision-making procedures and the future of the Plains. Sitting there, I sensed how depletion piggybacks on other legacies of exclusion. I wanted to know how my family was complicit.

I asked Susana and her son about racism. They did not shy away from the topic. They told stories about police and teachers at school. They said one teacher in particular put all the kids with Spanish names in the back and all the White kids in front. When Susana said she almost punched the woman, I believed her. I asked if she felt discriminated against by our family. Susana downplayed it. She said she noticed that Mexican workers got worse housing and tractors but that was it. There were problems, she said, but they always started in town, not on the farm. On the farm everybody had to work together. It's not like that in town. Everyone knows the cops are racists, even the Mexican cops. The jail is full of people with black hair, she said. What does that mean? It's like White people never do anything wrong and all the Mexicans are criminals.

The criminality that struck her family was White. The family was living up on the blacktop by the feedlot, still working for RW. The oldest son was eleven or twelve. He liked to play with his little brother. They made a cart. The older brother pulled it with a rope while the younger brother rode. One evening, they were going back and forth across the blacktop in the cart. A car was heading down the road fast. Her son pulled the cart to cross back towards the house but the wheels caught on the edge and the rope pulled off and the cart flipped. The boy who was pulling stumbled into the middle of the road. The car swerved to miss him and hit the cart with the young child dead center. The small boy died right there in the ditch. Two White people from Colorado were driving. They were friends of a prominent family in town and were heading home after a visit. They had been drinking. Susana went with the White people to RW's house and called the sheriff. The sheriff came but so did the local friend of the people who had killed the boy with their car. You know, Susana said, they didn't do anything to the people who hit my kid.

There was nothing real I could say that would not sound trite. Susana changed the topic. After another half hour of small talk, we stood up to go. Susana asked her son to take a picture of her standing on tiptoes between us in her kitchen. I told her she was welcome to come out to the Little Rock House whenever she wanted, that my father would grill out and play the guitar. She said it was sometimes hard to find gas money but she said she thought about the place often. I'd like to go, she said. Maybe someday.

* * *

The next day we bunched up the cattle. Five yearlings were missing. My father knew which ones just by looking.

It's that damnfool big steer again. It's always trying to break out. Like a teenager. It's the leader and the other four follow it. The problem is that it's so dumb.

We walked into the sand hills. Mostly any missing cattle were just concealed behind the first ridge. They were not there. We looked down the riverbed. No sign. Back at the corrals, we saddled the barn-soured horses and rode the north pasture. The yearlings were not in the draw or the ravine or the corner by the grass or on east. We didn't see any tracks or broken fence lines.

Without water, the animals wouldn't last long in this heat. Three days, maybe four. He wondered if they would come back that evening for a drink. They didn't show. The next morning he was up before dawn. We rode back down the river in the cool morning. After two hours of checking the fences again we found one weak segment in the far south corner. The fence was standing but could have bent down if they leaned on it. Twenty yards on the other side of the low fence he found half a hoofprint in a spot of dirt. He stopped and looked around for a long while.

I bet they went up onto that grass.

We rode in that direction. Just at the place he'd thought, we found fresh tracks in the road. In the night the animals had come back to the gate, desperate for water. He guessed they must have doubled back across the hillside and wandered onto the flats. There was nothing to stop them for miles and no water either. We rode back to the house and turned the arthritic horses out and got into the truck. Four miles northwest, he saw the yearling cattle plodding through stubble. We drove the truck over the furrows. When we got close enough for them to hear, he got into the back and started shaking cubes in a bucket. They lifted their heads at the familiar sound of his voice and the rattle of the feed. One called out. They trudged slowly toward the truck, one's tongue hanging out of its mouth. I idled the truck towards the pasture as my father walked behind, shaking the bucket until he was able to let them back in the fence. They knew their way to the tanks where the other cattle waited and watched them come. There, they drank deeply.

We bunched the yearlings with the others. After a while, we ran them all through the clanking chute. The red cattle were fat and gentle. It felt like they were doing us a favor by not kicking the whole thing down. My dad worked the levers and did the doctoring at the same time. My job was to get them in the alley and load them with the hotshot. The back bar was a six-foot iron pole. I kept putting it in the wrong place. Too high or too low, too close or too far. I crowded the cows or gave them jumping room or space to hurt their hocks. It caused more work for my father. With a competent partner he'd have been done in half the time with half the risk to the animals. But he never complained or said anything critical about my mistakes. He just gave me little tips. All he said was bring her up. Or, drop it down a little, Luke. Each cow got a soft slap when he let her go.

Thanks, he said afterwards. I couldn't have done it without your help.

I looked sideways at him. He kept a straight face.

I smiled. Any time.

We'd worked side by side all day. Afterwards, we hung on the fence. The wind cooled our sweat. Most of the cattle were grazing east of the house, headed back to the riverbed. The dogs lapped water. Barn swallows twisted and glided. I looked at the Little Rock House and recalled a photograph that Fern had taken in the late 1940s. My grandfather sat on a tractor, peering at Fern from under a short-brimmed hat. My father, an infant, was astride his leg. Behind them were the thick concrete walls

FIGURE 25

of the house, old even then. For a long time, I thought it was the mystery of destruction alone that was calling me back and that it had nothing to do with fathers or sons.

My father and I rested in quiet against the rusty wire. It felt calm. The past could not be rectified. There were things that we had done that could not be corrected, damage we had caused that could not be repaired. We had both lost some chances forever. Neither of us could outrun our mistakes or the consequences of our failures. As I stood there in silence, I knew then it was my father who had opened space for me to return. I had needed him, after all. And he had not disappointed. Somewhere along the way, he had turned our shared concern with the aquifer into a new kind of trust. It allowed us to have conversations that were not possible in the past and that would not always be possible in the future. It was no redemption, not for him and not for me. But it was a lot more than we'd had. That day, it was enough.

I looked west over the red grass where the evening sun emerged from a bank of clouds. I watched the shapes as they

FIGURE 26

billowed into columns and then collapsed into two bands of purple. Between them splayed fingers of light and the scene was cupped in the setting sun. Above the purple crests were lighter undulations and together they gave the impression of waves breaking upside-down on the shoreline of the horizon. I searched for a greater pattern but did not find a sign. Any story could be spun from the arrested motion figures. Standing there beside my father I wondered if a made-up parable with a clear ending might bring me closer to forgiveness or to the truth. I doubted it. The vaporous shapes reminded me how hard it is to know what is really solid in our relationships and what might vanish forever without special care.

The clouds formed every evening. Every night they rolled away. In the heat and the drought, farmers were pumping ever more water from the aquifer. Most of the moisture evaporated before it hit the ground.[18] The vapor ascended into the heavens where it appeared in these mythical shapes whose ultimate meaning eluded me. I realized that they were the negative image of the aquifer etched into the firmament and thought of the Old Testament verses that I had learned as a child on these Plains. I recalled the stories about the wells that Abraham drilled on the Negev Plain and the conflicts over the living waters of the desert that eventually pitted brother against brother and then I thought of the passages wherein were described the great flood unleashed at the beginning of the world when the waters of the deep merged with those of the skies to cleanse the surface of human corruption. I could not say how the ancient waters of the Plains weigh on some greater scale of justice or how its balance would tilt when they have run out. I do not know if there are mortal sins against the hand of the Creator or whether depletion can be judged apart from its consequences. The beauty and movement of the clouds as they drifted inexorably away should have terrified me. For some reason they did not.

* * *

We heard outlandish rumors about a court case over the aquifer in the neighboring county. The plaintiff, Jay, was one of the few other farmers who spoke out against depletion. The case had received national attention.[19] I was eager to meet him. Over several conversations from 2016 to 2018, he shared his perspective on justice and decline.

We first met at his office building in the small town where he was raised. The tall, square-jawed football star was married to his high-school sweetheart. He said he was pro-irrigation, pro-life, pro-Christian, and pro-business. A lifelong Republican, he had been appointed to the state Water Authority and two governors from different parties had appointed him to the State Board of Agriculture. Along with his brother, Jay runs a five-thousand-acre irrigation farm that has been in his family for four generations. He is a founding shareholder of a grain elevator, trucking company, and cotton gin about fifteen miles from the Little Rock House.

Jay said he learned about the dropping water table from his grandfather. After graduating with a degree in agricultural economics and finance in the mid-1990s, he returned to the farm. He and his brother vowed to be good stewards of the land. They set a goal of using 25 percent less water and still turning a profit. One way they did so was by introducing a kind of genetically modified cotton. It required one-third the amount of water as corn and sold for the same price. Cotton took off in western Kansas. For innovations like that, Jay and his family were named first runner-up for the 2004 "Producer of the Year" award given by *Top Producer* magazine.

I had rarely met anyone who so fully fit the image of a successful agribusinessman and local ideals of a good man. So I was surprised to learn that he felt ostracized in the town where he grew up. He told me he had gotten a series of death threats and thought his family might be in danger.

The conflict began in the early 2000s, when one of his wells went dry. In Kansas, water rights are allocated by the so-called doctrine of prior appropriation.[20] The first person to discover a resource can exploit it as private property. The doctrine follows a simple equation: first in time is first in right until the rights of others are impaired. If there is unlimited water, it works fine. If not, it creates problems.

Jay narrated the history of his legal case. All the case documents were publicly available; later I filled in the gaps.[21] I learned that his family's well was drilled in the 1930s and became a vested right in 1950. There are five neighboring wells within a two-mile radius, all drilled much later. State hydrologists discovered that the neighboring wells pumped an average of 1,500 acre-feet per year from 2005 to 2013, when water levels dropped so close to bedrock that Jay's well couldn't pump at all.

In 2005, Jay's family filed a complaint with the state Division of Water Resources, alleging that two junior water rights had impaired their senior vested rights. According to a letter written at the time, their goal in doing so was to bring attention to the urgent state of aquifer decline in general. In 2012, the neighboring property was sold to a company owned by a conservative billionaire whose son was a member of the Southwest GMD board. In spring that year, when farmers were flush with seven-dollar corn, Jay's family filed a lawsuit against the company for continuing to impair his water rights. It was the first western Kansas lawsuit over dry wells.

A series of motions and appeals followed. The Kansas Division of Water Resources issued a detailed report in 2016. It found that the aquifer had declined six feet per year in this area over the prior five years, owing to almost no recharge and extensive pumping. The report showed that Jay's ability to irrigate was prevented by his neighbors' water use. "If recent practices continue," the report noted, "well operators in the area faced the imminent end of the productive life of the aquifer."

Jay won the case. In 2017, the district court judge ruled that Jay's right was impaired by his neighbors. There was only enough water left to provide a limited supply to one well, the ruling noted, and even then, not for long. "It is apparent," the judge wrote, "that water is being depleted at a rate which cannot be recharged." The ruling described the situation as a "race to the bottom of the aquifer." As the race takes place "the only certainty in the system is to honor the system of first in time, first in right." The district court placed a permanent injunction on the neighboring wells. It ordered the company to cease pumping. Noting that legal mechanisms alone could not stop depletion, the court nonetheless ruled that giving senior water rights precedent "will foster certainty and may allow other remedies to slow the depletion of the aquifer."[22]

Outsiders hailed it as a historic victory for conservation. Jay's work was featured in newspapers and magazines across the country. The moment was ripe with promise. At first, Jay trusted the ruling would catalyze a larger intervention. Putting all junior water rights at risk provided an incentive to find a more equitable solution. He was assured by some that his lawsuit would give them the political cover necessary to take stronger actions to protect the aquifer. My goal, he told me later, was never to kill off junior water rights. It was to draw attention to a shared problem and push folks to come together for the future.

It could have been a turning point. But like some other historic possibilities, this one failed to materialize. Jay's support crumbled. He lost his position on the State Board of Agriculture and was disinvited from Farm Bureau meetings. Just as with Dwane, rumors were spread that Jay was out for his own selfish gain and was driving younger farmers out of business. I was naïve, Jay said. I thought people would embrace it. But the problem is, I tried to work within the system.

When I talked with him shortly after the ruling, he felt like it was a hollow victory. The favorable ruling was based on a state hydrology report that claimed Jay's and the five neighboring wells draw from a single, isolated pocket of groundwater. In other words, it presumed the decline in his water was not linked to wider practices but a micro-local problem. The pushback kept intensifying. Soon after the injunction was issued, certain lobbyists started pushing a bill that would oblige a farmer to exhaust all administrative remedies before a legal action could be filed. If successful, that would allow the GMD board and the state to stall on future lawsuits indefinitely. Around the same time, Jay's family started receiving those death threats. They were snubbed by lifelong friends and neighbors. In our attempt to be catalysts for change, he told me, we've become pariahs.

A year and a half after the ruling, I followed up with Jay. This time we met at his barn office. Built on the site of his great-grandfather's homestead, it was paneled with wood from the original barn and a roll of canvas off an old header. The case had gone into appeals. He had spent more than a million dollars of his own money on legal fees. His wife sometimes thought about moving because of all the gossip and snubs and the death threats.

By then, Jay had come to see depletion as a moral test for Plains society. At stake were his deepest held values and principles. He attributed much of the problem to what he called a water-mining mafia that manipulated the terms of water governance towards corporate interests and short-term profits. What was harder for him to understand was why so many good Christian people would go along with rapid depletion. He said that people he had known his entire life were increasingly committed to a hardcore water-mining mentality.

I've just been amazed at how entrenched the water-mining mafia mentality now is, he told me. Twenty-five years ago I'd say the water-mining mentality was 5–10 percent of the irrigators. Now I think it is much more. First people try to scare you, he said after a pause. Then they introduce this idea of futility. The water miners get bigger and bigger and say its futile to resist. And unless you have the burn in your eyes, the deep sense of what's right or wrong, that conviction of principle, you fold.

Jay's own sense of right and wrong came from his great-grandparents, Holdemann Mennonites and Missouri home-steaders who stuck through the Dust Bowl. His great-grandmother taught him that being a steward of the land was not a privilege. It was an obligation imparted to man with deep religious under-tones. Taking care of the land was a moral responsibility. Prior generations, she said, had sacrificed so that he could prosper. It was his duty to pass a sustainable way of life on to the future.

Jay had done his best to meet that obligation. But he had overextended to cover the legal bills. Then corn prices dropped. Hail took half his cotton crop. He confided that he might have to sell the farm. The battle for the aquifer had brought his family to the brink of ruin. You know, the funny thing, he concluded, is that to stay in this legal battle our family has pumped a higher amount of our legal water over the last two years than we did in the prior ten.

As I drove away from Jay's barn, I thought about how his trib-ulations stripped the story of decline to its bones. Community values entangled in a destructive system. Agriculture reliant on short-term extraction and toxic chemicals. Depletion justified in the name of illusory and absentee profit. Water governance aligned with corporate interests. Older notions of stewardship wither as a water-mining mentality spreads over the land. Good people give up and join in. Neighbor pitted against neighbor, generation against generation. A few decide to take a stand for what they believe is right, to hell with the consequences.

There are reasons to believe the aquifer under the Little Rock House will soon vanish forever. It may even be likely. Depletion is tied to family relations and habits that are passed down through generations.[23] But this extinction is not inevitable. The aquifer holds other stories alongside those of inexorable loss. Many farmers across the Ogallala country are not giving up. Jay, for one, found a different kind of ending. After his family won the final appeal of the court ruling in 2020, he changed his operation to grow hemp for oil, which yields one of the highest profits per unit of water of any crop that is possible to grow on the Plains. With the windfall, he was able to save his family's farm, for now. Others were not so fortunate.

After all my wanderings, I was left to question how we could really earn a second chance and what we would do with one if we had it. It is difficult to accept obligations to others. And it may be even more challenging to embrace mutual responsibility in the places where our original scar tissues were formed. Yet searching for ways to save the water means searching for kinship, real or imagined.

* * *

The summer lengthened. So did the drought. Every day the sun hung heavy over the plains. Every evening the cloudbanks formed and the heat lightning flashed without a drop of rain. By then, groundwater had so many stories to tell that it was hard to know what it said to each of us.

One hot afternoon I walked far down the Cimarron. Like always, my eye was drawn to the smallest details and the largest planes. What stretched between them was more difficult to grasp. Under the vast sky, the dogs moved as if synchronized. Two mule deer stared from a half mile away, ears splayed. I crossed the dry riverbed and climbed the sand hills, watching where I stepped. Cactus hid under the coarse grass and sagebrush.

When I crested the first ridge, I paused. From above came a faint trilling. After a long search, I spotted a ragged V of sandhill cranes, very high and almost invisible against the blanched sky. These cranes are among the oldest of birds and they follow ancient migration routes. They rarely pass this area. But those that do seem to remember prior versions of the land. Fern told me there was once a playa lake just north of the Little Rock House. Like the others, it was plowed decades ago. Still, the cranes always circle that spot before they continue. Were they mourning a loss or anticipating a change? I listened to their persistent calls. They made a wide loop and continued on.

The noise of a spray plane cut against the cranes. It moved in the opposite direction, flying much lower. From its number I knew it belonged to the same company that my father had reported several times for spray drift. He saw the pilots spraying chemicals on his land and watched the toxic fog settle on the grass. He moved the cattle and called the state. They sent an inspector, who took samples. The inspector found the grass contained high levels of lambda-cyhalothrin, an insecticide, and glyphosate, a herbicide. The company was issued a small fine.

At a half-buried fence line I turned northeast. Weeds— palmer amaranth and kochia—covered it almost to the top. Most native plants had gone dormant in the drought. Not the weeds. They were five feet tall, a spiky green. Agronomists recently discovered that some populations of these species have become resistant to the most toxic herbicides.[24] Like the flies that once traveled with bison herds and now swarmed on feedlots, their metabolisms outpaced the chemicals.

The shelf of sand above the riverbed was dotted by a handful of shallow pits where they once dug gravel. I skirted the pits and saw a metal stove cut from an oil drum and a ceramic dipper and just beyond the rusted arms of a machine that RW built in the 1950s. They called it the Farmhand. It used an inline six to power a front-end loader and a wire winch and was left in the river to disintegrate.

FIGURE 27

As I ran my hand along the pitted metal and cracked rubber, I recalled a story I heard about the end of RW's life. He turned everything over to his son. But RW had plowed so long he couldn't quit. Every day he asked what they needed him to do. They would send him to plow somewhere. Off he would go on his big Steiger tractor that bent in the middle. Sometimes the old man plowed the same ground two or three times just to have something to do. One day, when he was more than eighty years old, they found him on the Big Circle. He'd been disking a field that was already disked and ran into a telephone pole. He snapped off the pole and broke five or six disks and was trailing a quarter mile of fence wire behind him. After that they took away his tractor.

Now the machine is junk and RW's bones are dust. His beliefs cast a shadow that outlived him. They were etched into the land I walked. Was my generation bound to pass them on? I do not know, then or now, how to break the cycle. But trying to accept responsibility for the future was the only way I knew to start. I thought of the farmers I had met. I recalled the range of perspectives they voiced. I wondered what kind of Plains world we will leave to those who follow. Will it include common ground and will aquifer waters still flow beneath it? Will the children of the future grapple with their absence? Would any of those children be mine?

Over the sound of my boot steps, I heard a coyote wail and the yammering of her pups as they tried to imitate their mother's note. As I listened, I noticed the drone of irrigation motors guttering in the breeze. The motors must have been running constantly for the last two or three months. With a start, I realized they no longer distracted me with their sound. How easily I had gotten used to it, too.

Lost in such thoughts, I had wandered directly opposite the Springs. The sun was setting. The dogs had given up and turned back. I walked past the line of bleached trunks and crossed the riverbed. Once the waters boiled cold through this very sand and trickled into dappled pools where lived otters and beavers and silvery fish that no one would see again. I climbed the bank to the shelf of the vanished Springs where the memorial sat. The exact spot was lost when the river flooded, shifted, and dried up. No matter. I circled the tombstone monuments then turned my back to them.

Facing west, I scuffed off a sitting place in the trampled short-grass. On the ground lay a small arrowhead of pink and white flint. I stepped around it carefully, taking care not to touch it. The last of the light was slanting through clouds building on the horizon beyond the irrigated circle covered with grazing cattle. For a brief moment, before the sun sank below the earth, the same amber light covered everything in the valley. The light

erased the undulations of the land and blurred its lines. Sprinkler sand hill weed crop fence rut bone furrow coyote bird cow
airplane motor: all merged into a single glowing mass with no
more substance than the sky.

I watched the scene disintegrate as darkness settled over the
valley. Fern was gone but she still reached out to me and her
words had brought me back to that place where my father had
given me new stories to call my own. Under me moved the ancient waters and in those waters were written all possibilities. I
lay there in the dark, under the timeless stars, my back pressed
flat into the cooling earth.

* * *

It rained my last night on the farm, the first rain in months. The
waters rode a wild wind with split lightning and booming thun-
der. My father woke me an hour before dawn, dressed in boots
and hat and bathrobe. He made coffee while I packed. Then he
offered to follow me the muddy nine miles to blacktop, to make
sure I didn't get stuck.

The rain left puddles in the road and they shone like mirrors
under a waning moon. Dozens of little owls fluttered up in the
yellow headlights. At the edge of mud and asphalt, my father
parked the truck next to my car. We stood beside the rumbling
diesel, looking out over the blue plain.

Hundreds of flashing white lights, each marking the end of a
pivot system, circled the horizon. They blinked like stars over
the plowed uplands and the dry riverbed and the lost Springs
and bygone times real or made up. I knew I would come back.
We watched the lights.

Imagine all of those gallons being pumped right now as we're
standing here, he said. He shook my hand good-bye. He opened

the truck door then paused. You know, I think Fern'd want you to tell it like it is.

He drove the truck slowly down the road, back to the Little Rock House. Above me were clouds. In the clouds the waters of the depths mingled with those of the skies. They drifted to destinations still unknown. I got into my car and headed east.

AFTERWORD

I did not grow up with the feeling that something was being kept from me.[1] No wisp of mystery hung over the straight furrows and vast fields that made up the backdrop of my early life. No doubts drove me to attempt to reconstitute the ties that bound the surface of the present so tightly to the eradications of the past or to the declines of the future. Like others, I was able to confine to an alien world the layers of destruction that underlay my formative years. To me, the dry rivers seemed entirely separate from the pains of growing up and from the business of irrigation farming.

Only much later, after all my wanderings, did I return to the Plains to discover traces of ongoing decline everywhere I turned. Once I caught a glimpse of them, they were all that I could see. I know now that it was the draining of the aquifer that underwrote the world of my adolescence and its most pivotal events. The sound of motors, the drift of toxic fogs, the odors of a meat-packing plant all pervaded the lessons I learned about what it meant to shoulder a man's burdens—lessons that I carry with me still, for better and for worse. Now when I see photographs of stacked bison skulls or dust storms, they blur into images remembered from my life and I cannot help but recognize something of myself in them. Looking back, I often feel as if I am somehow the offspring of depletion, as if it is inscribed into my bones in ways from which I will never entirely escape.

So where can a true reckoning with depletion begin and where does it end? With a strategy to update management practices through more precise forms of modeling and expertise? With the innovation of more-efficient irrigation technology and crop varieties that require less water? With a sociology that details how agrarian capitalism drains water and wealth from the Plains to enrich investors elsewhere? With a diagnosis of how this case illustrates White supremacy, toxic masculinity, or the sentiments and logics of settler colonialism? With a chart of the ways aquifer loss combines with climate change to make ours an era of planetary ends? With an optimism that things aren't really as bad as they seem?

Such approaches offer important insight. None of them is enough. Like ecological tipping points or global pandemics or deepening inequities, depletion is hard to pin down. The drive to break it into separate elements and then to rationally explain each one is a central motor of the problem itself. Addressing one of its causal elements allows others to proliferate unchecked. Hydrogeologic models, resource management guidelines, and regulatory frameworks may prevent anyone from grasping the real scale of the problem. The risk is that limited explanations hide the chaos of agribusiness under a haze of order and reason.[2] Such analyses cannot account for the ways the roots and branches of the issue accrete over generations into a kind of historical consciousness, much less offer effective solutions to its present operations. To complicate it further, these partial portrayals, whether critical or legitimizing, are all-too-readily politicized and then coopted into wider efforts to divide rural people from others, from themselves, and from a collective future.

My journey started off in exactly these misguided directions. But it led me somewhere else. Over and over, my travels brought me face to face with the way depletion pervades the possibility of my existence no less than the lines of crops, the shape of clouds, the greenness of plants, the chips of flint, the obsolete machinery, the absence of extinct creatures, the marginalia of my ancestors, the jargon of mismanagement, the magic of the bottom line, the half-remembered dreams, the durable attitudes, the suspicious numbers, the family secrets, the chemical sprays, the unspeakable events, the arrogant faith, the frontier masks, or the ways of holding on and letting go. At times, these layers collapse and fuse. At others, they are broken and opposed. They always appear as something else. I hoped that conjuring their pattern of blockage and flow might give me a chance to revisit past choices and somehow loosen the grip that depletion has on me and others.

Easier said than done. I was not prepared for the ways my language of decline was already depleted, too. I had few words to account for depletion as a personal dilemma as well as a technical and scientific one. Yet, for me, intimacy is where it begins and ends. Aquifer loss is not abstract. On the Plains, people do not have the luxury of confronting it piecemeal. The immense scale of depletion can be grasped only in the intimate registers by which people are forced to inhabit all of its contradictory elements at once. It runs through the connective tissues of everyday life, for some more than others. And it was often tangled up in the same vocabulary I initially relied on to disrupt it. I dared to return to the farm only when I thought I no longer needed anyone else. I presumed I was able to make the right decisions about what was entrusted to me. At the beginning, I could not see how this echoed the drive for profit, the conceit of control, or the self-absorbed individualism that artificially divides the aquifer into parcels of private property and allows a few to drain it at the expense of many. Reckoning with depletion requires more than a critical analysis of other people's mistakes. It means reflecting on the collusions that allow me to keep running out, too.

* * *

Plains people ended up as my teachers and guides in this search, rather than the other way around. I already knew that rural people did not conform to many stereotypes of them. But I learned that farmers did not agree on what depletion implied. Some were water miners who favored pumping out the last of the aquifer waters for their own selfish gain, and they were not afraid to admit it. Others were more ambivalent. Realizing they were trapped in an unjust system that compelled them to keep irrigating, they felt they had no other real choice. A few, like Jay and Dwane, tried to pioneer new ways ahead. All of their courageous efforts were met with hostility, insult, and sabotage by those who took their creativity and resolve as threats. Like me, they found themselves caught up in a web of contradictions, where reason often resembled insanity and where good intentions often turned into the effects they aimed to oppose.

The stories they told never quite fit within what I could safely say about depletion and groundwater. What I initially thought were answers began to appear like questions, lines like circles, and explanations like myths. Perhaps more important were the silences, the things that we could not talk about at all. About halfway through the project, I started to see how these emotional and epistemic fault lines mirrored my own. A version of the same values that justified depletion drove my initial critiques of it and could also catalyze a movement for change. Among other things, that meant the situation was never quite hopeless. If a blanket of irrevocable loss covered the Plains, a thread of human decency ran through and against it. I found this, too, across southwest Kansas, in the homes of those who shared their stories, in the memories of past kindnesses, and in the voices of those who refused to knuckle under. I felt it more closely in the recharge of my father's support and in the ghostly presence of Fern that gradually came to haunt my every turn. The chances she was denied and the ways she refused to give up urged me beyond despair, even as I learned that I could never really put myself in her place. Almost always, the stories I found were interrupted. Usually, they were organized around absence.

The gaps in what it is possible to say or to remember communicate something vital about running out. To approximate it requires a different genre of writing about destruction, one that can account for depletion's dead ends, buried sediments, and mirage-like qualities. Inspired by Fern, I tried to write as close as possible to the overlooked textures of depletion in order to unsettle its presumed form. This turned into writing in loops and layers that follow depletion all the way down. Such writing is irreverent to orthodoxies of any kind, including the desire to blame either rural conservatives or progressive elites for all of the systemic problems of this country. It is hostile to approaches that privilege esoteric theory from a distance over conceptual alliance with the experiences of real people trying to make sense of endpoints at once imminent and retreating.

Writing in such ways about water gave me another rickety way back to the place where my life began. It let me revisit lost moments and turn them over to uncover meanings that were not apparent in real time. I shared an early draft of the text with my father and my uncle. It seemed to break something open between us. It stirred up old misunderstandings and offered new resolutions. Reading it gave my father a chance to ask me questions, too. We were able to find a shared purpose that was not possible before. It gives me something to keep and to cultivate in the years ahead. Call it out for what it is, they told me, and let the chips fall where they may.

* * *

This account of depletion aims to oppose the managerial genres that perpetuate groundwater loss in the name of preventing it.[3] I learned that the regional water management board was a primary agent of depletion. It served the interests of an elite corporate few at the expense of everyone else. Intentionally or not, it acted to block certain grassroots efforts to slow decline. The apparent self-dealing of this institution was hidden behind obtuse jargon and the promise of pipeline dreams. Some state officials said they deferred to this board's policy of endorsed decline out of respect for law, democracy, and local values, even while others tried to hold water abusers accountable by checking meters and levying fines. This created a magical kind of reality, where collective behaviors were somehow at odds with institutional understandings of them. It was organized around a fundamental exclusion. In its current form, regional water governance is a form of pay-to-play democracy, reserved for the already privileged. Only those who own water rights can participate in meetings and vote in elections. That gives corporate water users outsized influence. It also means that the vast majority of citizens in southwest Kansas, including most townspeople and laborers, are excluded from the decision-making processes that will determine the fate of the aquifer upon which their lands and livelihoods depend. Water governance is not inclusive or open. Instead, it can reinforce the hierarchies of race, class, and gender used to separate those who can properly feel at home on these Plains from all those who cannot.

At this point, depletion comes back into focus as one of the wider movements that erode democracy, divide us from one another, and threaten to make exiles of us all. There is no denying the ways depletion can mesmerize us with its power to make fantastical fears come true and realize apocalyptic beliefs. It thrives alongside greed and hopelessness—sometimes as symptom, sometimes as cause. On the Plains, many folks have good reasons to be cynical. Depletion intensifies and creates parasitic trends that target certain people's lives. These include: the displacement of small farmers; the rise of supply chains that link unsustainable crops to industrial slaughterhouses, feedlots, dairies, and ethanol plants; the consolidation of a racially marked underclass of disenfranchised workers who perform the most dangerous jobs; a financial system that compels irrigated agriculture and siphons off the profits; the pressure points that allow corporate lobbyists to brand toxic depletion as the best expression of a rural way of life and to evacuate conservation tools; the manipulation of truth and erosion of public trust by right-wing media alongside the defunding of public education; the creation of rural ghettos; and the spike in so-called deaths of despair by suicide or substance abuse among downwardly mobile rural folks who have nowhere else to turn.[4] Is it hard to understand why, under such conditions, some people conclude that self-extinction is the closest they can come to dignity and liberation?

Few deny that this portion of the Ogallala aquifer is running out. Among the most surprising discoveries of my fieldwork was that some practical tools to slow depletion already exist. There are several policy steps that could be taken immediately. The state could implement a district-wide intensive groundwater use control area, or IGUCA, which allows it to cap withdrawals at sustainable levels. The laws and mechanisms to do so are in place. If that is too daunting, the state could be asked to restrict irrigation to the actual water requirements of a crop or encourage farmers to grow crops with lower water requirements. If the state refuses to act, farmers could organize a district-wide LEMA, modeled on the one in northwest Kansas, or even a regional cap-and-trade system to minimize loss. Regardless, citizens should be empowered to vote on water management decisions. This might dilute corporate influence, transform the board and make its members more accountable to their own communities. Such seemingly small and commonsense steps would have big impacts. Examples of success can be found across the Ogallala region, where farmers from Nebraska to Texas are organizing and leading related efforts to slow decline.

If practical solutions exist, why aren't they being used? Cynicism, collusion, and exclusion are major parts of the answer. But all too often, it is also due to the ways the problem is understood by the institutions charged with solving it. Institutional focus on rational terms can overlook the complex ways people inhabit the structured intimacies of depletion, although this is the scale on which depletion is lived and can be effectively confronted. Even the best of existing policy solutions do not call for democratizing groundwater management or address the indirect injuries of depletion. They do not question how the pursuit of profit seems to hold an overwhelming allure; one that exceeds its actual conditions of possibility on the Plains. A strange momentum makes it hard for everyone to stop chasing artificial growth even when we know it is malignant or cannibalistic. Going against this grain requires a broader understanding of profit and a more inclusive sense of responsibility.

* * *

Responsibility, like depletion, is less obvious than it seems. When I started this project, I thought I knew who was to blame for aquifer loss. I must have somehow believed part of the scripts that separate duped, self-destructive victims from enlightened, cosmopolitan saviors. And I still cannot forgive those who direct and benefit from the eradication of the last remaining aquifer waters. But blame and self-grievance get us only so far, politically and personally. The overly assured sense that I could parse the guilty from the innocent was part of the problem. Like popular diagnostics of what is the matter with Kansas, this habit prevented me from grappling with fundamental questions about complicity. It allowed me to deny my role in depletion. It distracted me from what I hold in common with those on the Plains and it let too many of us off the hook. Assigning blame is not the same as taking responsibility; it is often the opposite. Either everyone is in this together, or no one is.

It comes down to this. I did not pump any water or shoot any buffalo or spray any toxins. But I am responsible to some degree for all of that, just as I am for climate change and racial inequity and the widening of acceptable disregard.[5] My generation may not have set any of these things in motion, but the momentum of destruction is carrying us forward at breakneck speeds. At a moment when this headlong motion seems poised to crash into dust, the question is whether we collude with the privilege and legacies of depletion or try to confront them. Rather than absolve me of responsibility, the recognition that I did not create depletion and cannot control it only increases my obligation to bear witness. Unless we scrutinize the long arc of depletion, we become a medium by which it is transferred to the future. And what we are undeniably responsible for is the future world we are creating, right now, all of us together. In many places, it is a world of running out.

When I look over the Plains, I sometimes feel as if I can glimpse what was once there and now is absent. Scenes of a vanished world and ghostly harvests and unrealized potentials float like mirages. They leave me dizzy and nauseous. Writing this account allowed me to reach out to my ancestors, to excavate their choices, and to locate myself within the remnants of their lives. So will the children of the future stand in judgment on me and you. I take it for granted that they will judge us harshly. They will seek to account for what was taken from them and what was left behind. But so often the same topics that get closed off from examination today are closed off from memory tomorrow. If that cycle is not interrupted, our reckoning with depletion and responsibility will remain incomplete. I offer this flawed testament in the hopes that future generations will have a fuller portrait of how one fragment of their world came to be.

I hope it may also inspire some on the Plains to take action now. Much can still be done. We stand at a crossroads. In one direction lies the final eradication of the aquifer waters. In the other is a chance to share some of this ancient life force with the future. We know that practical alternatives to depletion abound. They lie on collective ground. The choice is ours to make. Time to do so is running out. If we fail to act, an emptied aquifer will change our lives for us. Whatever we decide, it will reverberate far beyond this time and place.

* * *

Depletion is not limited to the details of aquifer loss in southwest Kansas. As the planet warms and droughts spread, similar dramas of aquifer loss are unfolding in dry areas around the world. The extreme losses of the High Plains are mirrored in the North China Plain, the Indus Basin of northern India and Pakistan, central Mexico, the Arabian aquifer system in Saudi Arabia and Yemen, the Murzuk-Djado Basin in northern Africa, and California's Central Valley. These foreshadow the declines underway elsewhere, including in parts of Australia, Israel, Jordan, Syria, South Africa, Namibia, Turkey, Bangladesh, Nepal, Chile, and Argentina.[6]

In each of these places, the meanings and forces of depletion are particular. They are inseparable from local genealogies and the lived intimacies of decline. They are also shaped by the characters of the dwindling aquifers themselves. Shallow, soft aquifers collapse when they are drained, causing the ground to sink under megacities like Bangkok and Mexico City.[7] Overpumping coastal aquifers in North Africa and the Middle East has reversed the flow of groundwater into the sea and turned aquifers saline.[8] Shallow aquifers around the world are polluted by chemicals, while deep fossil waters like those of the High Plains recharge slowly or not at all.[9]

At the same time, these zones of depletion bear striking resemblances to one another. Many share related histories and climates, as well as the techniques of irrigation. Technologies developed to drain the High Plains were spread around the world as part of Cold War geopolitics and development schemes. It is no coincidence that many regions with the highest rates of decline were also the sites of these early efforts to spark a Green Revolution.[10] Now, extreme depletion has become a generative force that creates structural adjustments of a different order. It reshapes lands into its own image, creates similar textures of extraction, and generalizes the experiences of groundwater loss. This allows for unexpected kinds of kinship to emerge between far-flung people and places.

Southwest Kansas farmers might recognize the Wadi Fatima area of Saudi Arabia, a conservative place where the most prosperous farmers choose to overwater their crops, where 70 percent of groundwater is lost to evaporation and inefficient delivery, and where 354 of 360 natural springs had gone dry by 1991.[11] Likewise, Kansas farmers may see themselves in the race to the bottom of the Kuchlagh aquifer in the Baluchistan province of Pakistan, where the shift to water-intensive commodity crops brought deep wells. In the absence of attempts to regulate water use, poor farmers lost their land and the rich continued to pump until the Kuchlagh aquifer was effectively exhausted by the early 2000s.[12] Kansas farmers might encounter more of their kindred in the Hai River Basin of northern China, where the richest wheat farmers have exhausted the deep aquifers, or in north Gujarat, where groundwater levels are dropping by around twenty feet each year, or in Tamil Nadu, where 95 percent of the wells owned by small farmers have gone dry and irrigated acreage has fallen by more than half since the mid-1990s.[13] They may also find themselves reflected in Yemen, where the eradication of aquifers contributed to the outbreak of civil war and deepened one of the biggest humanitarian crises of recent years.[14] The list goes on and on.

Comparative accounts of depletion remain to be written. But the most extreme manifestations may share several general features. Depletion is a kind of "self-devouring growth" that exerts organizational pressure on the politics of legitimate life.[15] It flourishes wherever people inhabit the residues of settler invasions and forgotten genocides, traces of destroyed ecosystems, surges of boom-bust despair and simmering resentment, chemical disruptions and the specter of more heat and drought.[16] In these zones, similar histories and technologies coincide with scarred landscapes and ideologies of unceasing productivity and profit that easily blur into militant fundamentalisms when they collapse. Extreme depletion reflects and reorganizes the causal relationships between these seemingly disparate processes. It instantiates affective states and collective dispositions that are environmental, historical, political, economic, interpretive, aesthetic, embodied, and broadly comparable, all at once. The Sand Creek massacre and genocidal campaigns in western Kansas targeting Kiowa, Cheyenne, Comanche, and Plains Apache reverberate in the dispossession of Rankülche peoples in the semiarid Argentine Pampas or the massacres of Herrero and San in eastern Namibia, where aquifer losses are similar to those on the Plains. It is no coincidence that the fate of the Ogallala aquifer is revisited upon aquifers named the Puelches and Guaraní, or that the dry wells of the Arabian Peninsula mirror Wagonbed Springs and the sandy course of the Cimarron is retraced in the Río Desaguadero of central Argentina. Groundwater management in many of these places repeats the same autocratic doublespeak, while water pipelines and artificial recharge schemes proliferate in depleted areas around the world and act to amplify the techno-corporate marketing of magic-bullet solutions, available for the right price. This time, already-depleted people and places are asked to pay the ever-more expensive bill.

As I picture the dry margins of these foreign lands, I cannot help but feel as if I am somehow coming home. In my mind's eye I roam the desert sands and burnt stubbles of places I have never been only to meet a cast of inexplicably familiar people. A cry of recognition rises in my throat for the faceless siblings who surely await among the other children of depletion, no matter their color or creed or culpability. There are millions of us, the inhabitants of depletion now forced to reckon with the sins and absences of our forefathers that are inscribed into our bodies and lands, as well as the sharp inequities of loss. Our ranks are growing. The unnamed bond we share cannot erase all hierarchies but perhaps it may cross some of the hardening divides that others have drawn between us. We have entered a pact with the future that is not of our own choosing. It is sealed with the ancient waters that will never come back and it may be as thick as blood. I am left to wonder whether any of you might recognize something of your lives reflected in the stories of mine.

* * *

Drying aquifers hold several different possibilities. If addressed within existing terms, their continuing eradication is likely to intensify conflicts, exacerbate inequalities, and entrench fundamentalisms. Yet their imminent loss may also spur us to imagine new footings for shared ground within the broken worlds we have unevenly made and inherited. Depletionist industry thrives when dignity and rights artificially appear as if they were zero-sum games, like nonrenewable resources to be selfishly hoarded before they run out. All too often, we are reduced to the crudest essences of race, class, gender, faith, or origin, and then convinced we do not share any common goals with one another. The result is that many of us deny our role in a social collective and refuse to privilege wider well-being over our own private gain. It is no coincidence that the bases for collective action are eroding just as impending planetary crises make the need for them increasingly clear. At a moment of rising hate, inequality, and essentialism, hope for a sustainable society lies in precisely the opposite direction.

Aquifers insist on a kind of overarching interdependence, whether we admit it or not. Their expansive natures push those who reside on depleted lands to acknowledge a common, species-level fate and turn its imperatives against the forces that seek to pit us against one another.[17] Doing so implies several steps. It requires building broad alliances focused on issues of unifying concern. It means using those alliances to effectively confront the structures, habits, and agents that are irreversibly destroying natural resources. It implies reestablishing shared criteria of truth, trust, evidence, and causation. It demands fostering an inclusive sense of mutual obligation that is not blind to the legacies of dispossession, past and future.[18] And it means thinking more seriously about how to build collectives around responsibilities that extend beyond one's own time, place, and self-identified group. If brought into a comparative framework, such steps may help to locate kinship within exile, community within exclusion, and alliance within disregard.

Ethnographic fieldwork, the central tool of anthropology, offers a method that could be suited to an antidepletion project.[19] It reveals patterns where there appear to be none and turns intolerable conundrums into questions amenable to analysis. Turning this method back on myself, my family, and the structured absences of Plains society changed my understanding of a problem that otherwise seemed paralyzing. It taught me that the immense scales of depletion are best understood through their most intimate fragments, and that trying to respond to a planetary crisis begins with a critical reckoning with the terms of my existence, complicit and otherwise. This method refuses to surrender the flawed present to past wrongs or to a future of inevitable demise. It opens space for unexpected kinds of recharge to occur. And it insists on people and horizons that remain open-ended, incomplete, and yet to be determined.[20] Such perspectives can help orient our search for a future beyond running out.

* * *

In the winter months, most of the irrigation motors fall silent. Farmers can't pump in the cold. Buried seeds wait for water and warmth. Quieter rhythms surface. The stubbled space opens to the ancient play of light and wind. I made it back to Wagonbed Springs fifteen months after I'd left. Harvests had come and gone. Another winter was on its way. The sky was gray. Dry snow covered the ground. It smoothed the yellow grass and filled the hollows. This time, I was not alone. My father came with me and so did Kate.

We got out of the truck and walked past the water faucet and the metal buffalo and the tombstone markers down to the sandy riverbed of the Cimarron. We wove between the fallen trunks of the cottonwoods and a bovine skeleton disarticulated by coyotes and the wind. I hung back and let them go ahead of me, the old man alongside the young woman. The wind drowned out most of their words. I thought I heard something about Fern and cattle and children and writing and coming back home but I wasn't sure. My father started to sing and she tried to add a harmony and then they were laughing. As I listened to the sounds blown out by the wind, I thought about the things that couldn't be undone and those that I still had time to do. I watched the two of them move together, the living among the dead, over the sands of a vanished river in a place haunted by waters where I never thought I might find another chance.

ACKNOWLEDGMENTS

A book this long in the making entails many debts. My greatest thanks are due to the people of southwest Kansas whose perspectives comprise the core of this work. Between 2016 and 2018, dozens of individuals and families graciously shared the expertise, insights, reflections, and observations that made this account possible. That most remain anonymous should only underscore the depth of their courage, generosity, kindness, and assistance. Conversations with Jay Garetson, Dwane Roth, and many others provided indispensable orientations and vital information. I learned much from experts and staff at Grant County offices, Haskell County offices, the Kansas Department of Agriculture, the Kansas Division of Water Resources, the Kansas Geological Survey, the Kansas Water Office, Kansas State University Extension's regional office, the Topeka Office of the USDA Risk Management Agency, Southwest Kansas Groundwater Management District Number 3, Northwest Kansas Groundwater Management District Number 4, the Kansas State Historical Society, the Historic Adobe Museum, Boot Hill Museum, and the organizers and participants of the 2018 Ogallala Aquifer Summit.

At the University of Oklahoma, my colleagues and staff in the Department of Anthropology provided the time, space, and support to pursue this project. I owe a special thanks to the students in my seminars on ethnographic writing at Oklahoma, whose talent, questions, and commitments set a high bar.

I was fortunate to draft most of the text while a 2018–19 fellow at Harvard's Radcliffe Institute for Advanced Studies. The environment and fellows at Radcliffe changed my approach to this project. I am grateful to all, including Robin Bernstein, Tomiko Brown-Nagin, Katie Bugyis, Stephanie DeGooyer, Moon Duchin, Malick Ghachem, Lauren Groff, EJ Hill, Sky Hopinka, Min Jin Lee, Charlotte Lloyd, Susan Meiselas, Christia Mercer, Durba Mitra, Nicole Nelson, Meredith Quinn, Janet Rich-Evans, Dana Sajdi, Evie Shockley, Karmina Silec, Fern Silva, Katherine Turk, Javier Zamora, and Chiara Zurzolo. Special thanks are owed to Jessica Bardsley, Tanisha Ford, Francisco Goldman, Stefan Helmreich, and Hernan del Valle for conversation, camaraderie, and feedback. Undergraduate research partners Julian Rauter and Sophie Westbrook diligently checked for errors, located sources, and contributed crucial information. Due to the vision, consideration, and stewardship of Fred Appel, the book found a home at Princeton University Press. Comments from two anonymous reviewers improved its final form, as did contributions from Maia Vaswani and Maria Whelan.

Many people have provided suggestions and critiques of various iterations of this material. The book benefitted enormously from engagements and feedback offered by Nikhil Anand, Andrea Ballestero, Anya Bernstein, Alex Blanchette, Viktor Coffey, Jennifer Deger, Les Field, Michael Fischer, Susanne Freidberg, Rivka Galchen, Faye Ginsburg, Stefan Helmreich, Kregg Hetherington, Melinda Hinkson, Robert Hitchcock, Jean Jackson, Ieva Jusionyte, Doug Kiel, Chelsy Kivland, Ryan Klataske, Misha Klein, Don Kulick, Marcel LaFlamme, Willi Lempert, Dan Mains, Kate Marshall, Emily Martin, Joseph

Masco, Fred Myers, Priya Nelson, Todd Nicewonger, Laura Ogden, Heather Paxson, Asa Randall, Elyse Singer, Michael Taussig, Binky Urban, Eve Vincent, Krista van Vleet, Chris Walley, Lezlie Ybarra, and Erin Yerby, among others. Lauren Groff's generous comments on an early draft shaped its final form and my understanding of craft. Francisco Goldman offered pivotal encouragement, reflections, and friendship that extend beyond the writing of this account. Gastón Gordillo's careful reading of the manuscript and our on-going dialogues about the generative dimensions of destruction were particularly important.

Some of my greatest intellectual debts are owed to close friends. Teresa Montoya helped me to better understand the moral economy of settler sentiments and toxic legacies. Laurence Ralph urged me to rethink ethnographic genres and taught me much about what it means to do anthropology now. Amy Moran-Thomas read each line more than once; she pushed me to dig deeper, to think harder, and to write more honestly every time. David Bond offered incisive comments that oriented my understandings of this material, our vocation, and the state of the world. Many others offered personal support along the way. This account would not have been possible without the steadfast alliance, intellect, and care of this larger community.

Likewise, the process of writing this book brought me closer to those I consider my extended family. Chamia Chiqueno, Chagabi Etacore, Chiri Etacore, and Erui Etacore taught me much about grit and resilience. Harald Prins has been a bedrock source of advice, courage, and solace in every step of this project and throughout my entire adult life. Bunny McBride first showed me the power of writing. She gave extensive comments on several drafts of the text and imparted wisdom that transcends it. Linda Bessire, Somer Bessire-Briers, and Jesse Bessire are present in every word and every breath. Mick Bessire, who knows more about the topic than I ever will, shared vital stories and wise guidance. Antony Glenn Bessire was a partner and collaborator in this from beginning to end. His care and observations gave me a way back and a path ahead. Towards the end of this project, I met Kate Haughey, who not only discussed every passage but whose presence illuminated horizons yet to come.

These pages cannot fully convey the ways my existence was transformed by the generosity, knowledge, consideration, and perspectives of so many. All shortcomings written or omitted, of course, are solely my own.

NOTES

Pseudonyms are used throughout the book. Certain details of scenes, characters, and places have been altered or omitted in an effort to protect the privacy of people, living and dead. The text draws from a large range of published and unpublished texts for numbers, statistics, and many other points of information. Citations of these pieces follow, and references can be found in full in the bibliography. I have not provided sources for facts that are not in dispute, considered common knowledge, or that can be easily found across many of the listed citations. Additional works of scholarship and literature have offered crucial frameworks and orientations for this account. The most influential of these are also in the bibliography.

Lines

1. Here and throughout the text, I draw upon a great deal of scholarship that examines water and its materialities as critical prompt, figure of thought, and sociopolitical imaginary. Perramond's *Unsettled Waters* and Worster's *Rivers of Empire* are particularly influential framings of these issues in the American West. See also: Soll, *Empire of Water*; Steinberg, *Nature Incorporated*. My thinking about groundwater is inspired by many anthropological, historical, and geographic analyses of the relationships between water and infrastructures, social organization, environmental justice, human rights claims, legal frameworks, governance, ascriptive hierarchies of difference, concept work and wider political economies. See especially Anand, *Hydraulic City*; Bakker, *Privatizing Water*; Ballestero, *Future History*; Barnes, *Cultivating the Nile*; Bjorkman, *Pipe Politics*; Gleick, "Human Right to Water"; Hastrup and Hastrup, *Waterworlds*; Helmreich, "Waves"; Johnston, "Political Ecology of Water"; Kane, *Rivers*

Meet the Sea; Kelly, "Study of Irrigation"; McKittrick, "Making Rain"; Mehta, *Politics and Poetics*; Muehlebach, "Price of Austerity"; Muehlmann, *Where the River Ends*; Ogden, *Swamplife*; Orlove and Caton, "Water Sustainability"; L. Roberts and Phillips, *Water, Creativity and Meaning*; Von Schnitzler, *Democracy's Infrastructure*; Wutich, "Water Insecurity"; Wutich and Beresford, "Economic Anthropology of Water."

2. Famiglietti, "Global Groundwater Crisis." See also Cruse et al., "Irrigation Aquifer Depletion"; Famiglietti et al., "Satellites Measure Recent Rates"; Gleeson et al., "Water Balance of Global Aquifers"; Scanlon et al., "Groundwater Depletion and Sustainability."

3. See Chatterjee et al., "Mapping and Management of Aquifers"; Feng et al., "Evaluation of Groundwater Depletion"; Huang et al., "Subregional-Scale Groundwater Depletion"; Rodell, Velicogna, and Famiglietti, "Satellite-Based Estimates of Groundwater."

4. Some hydrologists estimate that groundwater depletion accounts for as much as one-fourth of sea level rise: Wada et al., "Global Depletion of Groundwater Resources." See also Wada et al., "Past and Future Contribution"; Wada et al., "Fate of Water Pumped."

5. See maps at Kansas Department of Agriculture, "Southwest Kansas GMD Number 3." See also J. Butler et al., "Sustainability of Aquifers"; Kromm and White, *Groundwater Exploitation*; Whittemore, Butler, and Wilson, "Major Drivers of Water-level Declines"; and others.

6. Estimated recharge in areas of Grant and Haskell Counties in Kansas is 0.1–1.0 inches per year. See Kansas Department of Agriculture, *Final Report*, 3; Buchanan, Buddemeier, and Wilson, *High Plains Aquifer*. For more on the complex ties between depletion and recharge, see Sophocleous, "Groundwater Management Practices."

7. Hyndman and Hyndman, *Natural Hazards and Disasters*, 227.

8. Depletion is not uniform across the High Plains or the Ogallala aquifer. Some areas, mostly in the southern Plains, are effectively depleted, while others, mostly in the northern Plains, still have substantial reserves. This account focuses on southwest Kansas. While averages cannot tell the full story of aquifer loss, the mean decline in the twelve counties of southwest Kansas since predevelopment is around 103 feet. The area near the Little Rock House has one of the highest measured rates of aquifer decline in southwest Kansas. Measurements have shown this since the first recordings began in 1958. See Whittemore, Wilson, and Butler, *Status of the High Plains Aquifer*, 3. See also Kansas Geological Survey, "High Plains Aquifer Interactive Atlas."

9. Scanlon et al., "Groundwater Depletion and Sustainability." See also Peterson, Flynn, and Traylor, *Groundwater-Flow Model*.

10. Southwest Kansas Groundwater Management District Number 3, "Revised (2018) (Draft) Management Program."

11. Here and throughout the text, I draw inspiration from the critique of "agribusiness writing" and the anthropological alternatives to it developed by Michael Taussig. For an early formulation, see Taussig, "Corn-Wolf."

12. "An Ocean of Water," *Syracuse Journal*, February 23, 1906. See also Ashworth, *Ogallala Blue*.

13. Gutentag et al., "Geohydrology of the High Plains Aquifer."

14. Buchanan, Buddemeier, and Wilson, *High Plains Aquifer*.

15. Water witching strikes at the paradox of knowing and not knowing embedded in subterranean water and its defining mythologies. For more on this branch of sympathetic magic, see Hyman and Cohen, "Water-Witching." See also Kansas State Board of Agriculture, *Twenty-Third Biennial Report*, 75; Kilian, "Observations of the Divining Rod"; Krautwurst, "Water-Witching Modernist Epistemologies"; Lovewell, "'Witching' for Water," 103; Vogt and Hyman, *Water Witching, U.S.A.*; Vogt and Golde, "Folklore of Water Witching"; Waters, "Water Supply of the Farmhouse," 15.

16. Here and throughout the text, I draw inspiration from Gaston Gordillo's insights on the generative qualities of the rubble created by colonial violence, as well as Ann Stoler's analysis of imperial debris. See Gordillo, *Rubble*; Stoler, *Imperial Debris*.

17. The ties between western Kansas and the Paraguayan Gran Chaco run deep. This extends to theories of water and drought. James Espy, a nineteenth-century meteorologist whose work influenced rainmaking projects in Kansas, drew inspiration from an account by a missionary working in the Paraguayan Gran Chaco named Martin Dobritzhoffer. According to Espy, Dobritzhoffer observed smoke creating thunderstorms and attributed the natives' burning of the plains to an effort at producing rain. See Espy, *Philosophy of Storms*, 494. The cooperation between the two regions is formalized through the Kansas Paraguay Committee. Started in 1968 by the 501(c)(3) Partners of the Americas, it links the country of Paraguay and the state of Kansas to promote volunteering and social impact in both areas. See the website of the Comité Paraguay Kansas (Kansas Paraguay Committee) at www.cpk.org.py. For more on Mennonite environmental relationships in western Kansas, see Loewen, *Diaspora in the Countryside*.

18. L. Bessire, *Behold the Black Caiman*.

19. Here and throughout the text, I draw inspiration from a number of genre-crossing books that show how critical reflexivity, auto-ethnography, and family histories can be used to craft wider arguments about the relationship between the personal and the political or the environmental. Steedman, *Landscape for a Good Woman*, shaped my understanding of class, gender, and genre, as did two masterful accounts set in Kansas—Bair, *Ogallala Road*, and Smarsh, *Heartland*—as well as Agee and Evans, *Praise Famous Men*; Goldman, *Political Murder*; Hunt, *Underground*;

Stewart, *Ordinary Affects*. For some recent reflections on ethnographic writing, see Pandian and MacLean, *Crumpled Paper Boat*.

20. Here and throughout the text, I draw from the question of responsibility as framed in Baldwin and Mead, *Rap on Race*.

21. For a related point, see Z. Smith, "Elegy for a Country's Seasons." See also Galchen, "Weather Underground"; Orwell, *English Language*.

22. According to the 2019 draft of the GMD3 draft management plan there were 8,341 square miles and 4,400 farms in GMD3 counties as of 2012. See Southwest Kansas Groundwater Management District Number 3, "Revised (2018) (Draft) Management Program," 19, 27.

23. As of 2017, counties in GMD3 contributed about 33% of the market value of agricultural products sold in the state of Kansas: National Agricultural Statistics Service, *Kansas*; see also Southwest Kansas Groundwater Management District Number 3, "Revised (2018) (Draft) Management Program," 22.

24. Peck, "Kansas Groundwater Management Districts."

25. Groundwater Management District Act, Legislative Declaration, KSA § 82a-1020 (1972).

26. Groundwater Management District Act, District Powers; Home Office, KSA § 82a-1028 (1972).

27. Ramsey, "Kansas Groundwater Management Districts."

28. Sanderson and Frey, "Structural Impediments"; see also Brauer et al., "Ogallala Aquifer Program."

29. Southwest Kansas Groundwater Management District Number 3, "Revised (2018) (Draft) Management Program," 19.

30. The lower estimate of 0.1 inch for annual recharge is based on figures from the Kansas Division of Water Resources, while the higher estimate of 1.0 inch comes from the United States Geological Survey. See Kansas Department of Agriculture, *Final Report*, 3; see also Buchanan, Buddemeier, and Wilson, *High Plains Aquifer*. For discussion of global aquifer recharge in arid and semiarid regions, see Scanlon et al., "Global Synthesis of Groundwater Recharge."

31. Detailed discussion of the Cimarron River's hydrologic relationship to the High Plains aquifer can be found in H. O'Connor and McClain, *Ogallala Aquifer Study in Kansas*; see also Adams and Bergman, *Geohydrology of Alluvium and Terrace Deposits*.

32. Gregg, *Commerce of the Prairies*, 71.

33. Gregg, 79.

34. Wislizenus, *Memoir of a Tour*, 13.

35. From a letter dated 1941 by J. W. Dappert, a Grant County surveyor from Taylorville, Illinois, quoted in McLaughlin, "Accelerated Channel Erosion," 84.

36. Lucille Towler, interviewed by Fern Bessire, unpublished notes, February 22, 1968.

37. Dixon, *Life of "Billy" Dixon*, 117.

38. McLaughlin, "Accelerated Channel Erosion."

39. Kansas citizens have long been aware of the dangers that irrigation poses to the aquifer, as area newspapers tracked aquifer declines throughout the 1970s and 1980s: "Water Loss Problems Localized," *Hutchinson News*, June 17, 1976. For more such reports, see "Water Table Dropping in West Kansas," *Norton Daily Telegram*, January 26, 1967; "Groundwater Depletion Said State Danger," *Topeka Capital*, March 11, 1971; "Western Kansas Water Table Drops," *Topeka Capital*, March 7, 1975; "Kansas Irrigation Has Cloudy Future," *Kansas City Star*, March 20, 1977; "Kansas Wells Running Dry from Demands," *Kansas City Star*, November 26, 1978; "Water Is Slipping through Western Kansas Fingers," *Garden City Telegram*, June 24, 1980; "Study Criticizes Depletion of Kansas Aquifer," *Kansas City Star*, December 9, 1984.

40. VanLooy and Martin, "Channel and Vegetation Change." See also Schumm and Lichty, *Channel Widening and Flood-Plain Construction*.

41. "An Ocean of Water," *Syracuse Journal*, February 23, 1906.

42. Geologist Nelson Horatio Darton named these subterranean formations after Native groups of the Plains and is responsible for applying the label "Ogallala" to one of the hydrogeological layers. For early descriptions and observations, see Darton, *Geology and Water Resources* and *Geology and Underground Water Resources*.

43. "Irrigation May Open New Projects Here," *Grant County Pioneer*, October 22, 1931; see also "Brown Well Back to Use," *Ulysses News*, April 17, 1947.

44. Strong divisions between normative gender roles and gendered spaces persist on the Plains. According to regional cultures, agriculture is often perceived as a White, masculine public space. While women's contributions to family farming are well known and locally acknowledged, these heteronormative and racialized gender roles are patterned, delimited, standardized, enforced, and authorized in ways that merit further ethnographic attention. Scholars are working in that direction. For a sociological reflection on historical tensions of labor, economic diversification, and feminist resistance on Kansas farms, see Flora and Flora, "Women's Culture"; Schwieder and Fink, "Plains Women." For a discussion of more current gender roles among farming families in western Kansas, see De Wit, "Women's Sense of Place." See also Sarah Beach, "Views on Women"; Grant, *Down and Out*; Patterson-Black, "Women Homesteaders"; Sachs, *Gendered Fields*.

45. See "Self-Propelled Irrigation Works Wonders," *Wichita Eagle-Beacon*, October 17, 1965; see also Aucoin, "Irrigation Revolution," 17; Kepfield, "'Liquid Gold' Rush"; McKnight, "Great Circles on the Great Plains"; James Sherow, *Grasslands of the United States*, 312–13; Splinter, "Center-Pivot Irrigation."

46. In the mid 1990s, surveys indicated that more than 60% of respondents admitted to prewatering corn fields. The actual percentage is likely higher. See Stone et al., "Storage Efficiency of Preplant Irrigation"; see also Kisekka et al., "Optimizing Preplant Irrigation"; O'Brien et al., "Corn Yields and Profitability."

47. Stone et al., "Storage Efficiency of Preplant Irrigation"; see also Musick and Lamm, "Preplant Irrigation."

48. One study from northwestern Kansas found that 60%–80% of preirrigation water was wasted: see Kansas State University Agricultural Experiment Station and Cooperative Extension Service, *Corn Production Handbook* (1994), 27; See also Rogers, *Efficiencies and Water Losses*; Rogers and Lamm, "Soil Water Survey."

Bones

1. These figures represent general trends, not the specific experience of any particular farmer. According to the Kansas State University *Corn Management* guide, the statewide average irrigated corn cost of production was $3.60 per bushel in 2017. Multiplying by the 2017 average irrigated yield of 197 bushels/acre obtained from the National Agricultural Statistics Service gives a statewide average input of $709.20 per acre. In 2016, the statewide average production costs for irrigated corn were $3.96 per bushel, and $4.69 per bushel on average over the 2011–15 period. Data on regional trends would give a more precise picture, but much of this information is carefully guarded by farmers. See Ciampitti et al., *Kansas Corn Management 2019*.

2. According to Kansas Corn, the state currently has twelve ethanol plants, which create a yearly market for 183 million bushels of corn and sorghum. See Kansas Grain Sorghum Commission, "About Ethanol," and Kansas Corn, "Building Markets for Ethanol." For the social and environmental impacts of ethanol production in areas of groundwater depletion, as well as wider narratives of productivity that surround this industry, see also J. Brown et al., "Ethanol Plant Location and Intensification"; Condon, Klemick, and Wolverton, "Impacts of Ethanol Policy"; Drabik, Ciaian, and Pokrivčák, "Effect of Ethanol Policies"; Mishra and Yeh, "Life Cycle Water Consumption"; M. Roberts, Male, and Toombs, *Potential Impacts of Biofuels Expansion*; Selfa, Iaroi, and Burnham, "Promoting Ethanol in Rural Kansas"; Wu and Langpap, "Price and Welfare Effects."

3. A certain amount of annual debt is inevitable for any agricultural venture, so economists and bankers tend to measure the severity of debt by debt-to-asset ratio. These figures of debt percentage are based on data from the USDA Agricultural Resource Management Survey. Debt percentages from 2008 to 2017 show an average of about 26.5% of Plains farms are in debt. The average is 33.1% for Kansas farms and 47.8% for Kansas corn farms. See USDA Economic Research Service, "Tailored Reports"; see also Amy Bickel, "Are Farmers Successfully Managing Their Farm Debt?"

High Plains Journal, January 30, 2018; Kauffman and Kreitman, "Large-Scale Financing."

4. Davidson, *Broken Heartland.*

5. Barfield, *Rules and Regulations.*

6. USDA Risk Management Agency, *Corn: Kansas.*

7. USDA Risk Management Agency.

8. The structure of a two-tiered federal crop insurance could encourage further depletion in areas with declining aquifers. Recognizing this dynamic, in 2017 officials at the Kansas Regional Office of the US Department of Agriculture Risk Management Agency (USDA-RMA) pioneered an important third option for insuring irrigated crops grown under conditions of depletion. This is known as limited irrigation insurance. On changes to irrigation and crop insurance policies, see Kansas Water Office, "USDA Crop Insurance—Limited Irrigation."

9. See C. O'Connor and Bryant, "Covering Crops." See also Shawn Beach, *Irrigation and Limited Irrigation.*

10. This point has been examined in detail by many scholars. For more on crop insurance and its environmental impacts, see Adler, "Drought, Sustainability, and the Law"; Claassen, Langpap, and Wu, "Impacts of Federal Crop Insurance"; Deryugina and Konar, "Impacts of Crop Insurance"; Domina, "Federal Crop Insurance"; Kate Galbraith, "Texas Farmers Watered Crops, Knowing that They Wouldn't Grow," *Texas Tribune,* March 28, 2012; Walker, Singh, and Asokan, "Risk Benefits, Crop Insurance"; Weber, Key, and O'Donoghue, "Federal Crop Insurance."

11. As of the 2012 census, Grant County had 329 farms and Haskell County had 187, for a total of 516 farms. See USDA National Agricultural Statistics Service, *Grant County, Kansas* and *Haskell County, Kansas.*

12. This figure is based on data obtained from the USDA Risk Management Agency, which oversees the Federal Crop Insurance Corporation. The cumulative indemnity payments for Grant County from 2008 to 2018 were $81,369,583 and the cumulative payments for Haskell County for the same period were $74,324,618, for a total of $155,694,201. See USDA Risk Management Agency, "Crop Indemnity Maps Archive."

13. Sources agree on the timeline for the extermination of western Kansas bison. There is less agreement on the absolute scale of the slaughter owing to the difficulty of establishing the historical baseline size of the southern herd. For various estimates, see Dodge, *Hunting Grounds of the Great West,* 142; Flores, "Bison Ecology and Bison Diplomacy"; Hanner, "Government Response to the Buffalo."

14. The details of this period of settler colonialism are recounted in many sources, including Dixon, *Life of "Billy" Dixon,* 81; Gard, *Great Buffalo Hunt* and "Mooar Brothers, Buffalo Hunters," 34; Hanner, "Government Response to the Buffalo"; Mooar and Pace, *Buffalo Days.* For more on the circulars distributed to unemployed

track workers in western Kansas in 1871, see Joe Hutt, *Ellis County News*, November 5, 1925, quoted in Fleharty, *Wild Animals and Settlers*, 56. For an important exploration of the causal links between colonial violence and ecological changes during this time period in western Kansas, see Monnett, "Transitional Kansas Landscapes."

15. Mayer and Roth, *Buffalo Harvest*, 15–16.

16. Mayer and Roth, 21.

17. Dodge, *Hunting Grounds of the Great West*, 132.

18. Mayer and Roth, *Buffalo Harvest*, 18.

19. Chrisman, *Lost Trails of the Cimarron*, 20. See also G. Brown, *Life and Adventures*, 118.

20. Mayer and Roth, *Buffalo Harvest*, 35.

21. Mayer and Roth, 26.

22. Isenberg, *Destruction of the Bison*, 131.

23. Mayer and Roth, *Buffalo Harvest*, 41.

24. On so-called Texas beef, see Gard, *Great Buffalo Hunt*, 81. On smoked and pickled tongues, Mead, *Hunting and Trading*, 83. See also Mooar and Pace, *Buffalo Days*, 39; Mayer and Roth, *Buffalo Harvest*, 48; Fleharty, *Wild Animals and Settlers*, 48–53.

25. Cook, *Border and the Buffalo*, 101; see also Isenberg, *Destruction of the Bison*, 130–31.

26. Dodge, *Hunting Grounds of the Great West*, 133.

27. Dixon, *Life of "Billy" Dixon*, 81; see also Cook, *Border and the Buffalo*, 103.

28. Cook, *Border and the Buffalo*, 103.

29. For Indigenous place names, ancestral territories, and the Cimarron, see W. Clark, *Indian Sign Language*, 424; Grinnell, "Cheyenne Stream Names"; Meadows, *Kiowa Ethnogeography*, 279; Mooney, *Calendar History of the Kiowa*, 415.

30. Cozzens, *Earth Is Weeping*, 155–57; see also Gard, "Mooar Brothers, Buffalo Hunters," 36.

31. Dixon, *Life of "Billy" Dixon*, 94.

32. Mooar and Pace, *Buffalo Days*, 37. Also, Mooar, "First Buffalo Hunting," 109–10; Smits, "Frontier Army," 328.

33. W. Butler, *Sir William Butler*, 97.

34. For analyses of this historical paramilitary violence, settler genocide, and ethnic cleansing in the Panhandle country, see Anderson, *Conquest of Texas*; Chalfant, *Cheyennes at Dark Water Creek*; Hämäläinen, *Comanche Empire*. See Gilpin, *Central Gold Region*, for an example of common settler attitudes of the time. For analyses of the sentimental economies of colonial societies and the purification of difference, see Rifkin, *Beyond Settler Time*; Stoler, *Carnal Knowledge*.

35. Gard, *Great Buffalo Hunt*, 182–205, 217–55. See also Fouquet, "Buffalo Days."

36. This is a genre, as several newspapers published accounts of "the last buffalo" seen in various counties of Kansas. See "Last Buffalo in Jewell: One Killed on Mankato's Town Site in July 1873," *Topeka State Journal*, March 9, 1897; "The Last Buffalo Hunt in Stafford County," *Topeka Capital*, circa 1898; "When Buffalo Were Numerous in Kansas," *Hutchinson News*, October 29, 1874; "He Saw the Last Buffalo Killed," *Wichita Beacon*, October 10, 1926. See also Chrisman, *Lost Trails of the Cimarron*, 23. Newspaper clippings from this period quote several veteran buffalo hunters who show no remorse for their participation in the animal's extermination: "What an Old Buffalo Hunter Saw Who Helped to Exterminate the Herds that Darkened the Plains," *Kansas City Star*, November 30, 1930; "Old Buffalo Hunter Helped Clear Plains: Would Do It Again if There Were Bison in Kansas," *Topeka Capital*, December 20, 1931; "Dodge City Pioneer Saw Buffalo Disappear," *Topeka Capital*, November 8, 1931. One exception is Charles J. Jones, better known as "Buffalo" Jones, who regretted his participation in the buffalo's decline and worked to establish captive bison herds, as well as preserve other animals, such as muskox: "He Has a Scheme: 'Buffalo' Jones Wants to Establish Great Buffalo Ranch," *Kansas City Journal*, January 25, 1901; "Jones and His Buffaloes," *St. Joseph News*, November 1, 1902; "Of Buffalo Butchers: Railroad Man's Reminiscence of Early Days in Kansas," *Kansas City Journal*, circa 1901.

37. F. Bessire, *Glimpses of Grant County*, 13.

38. Sexson and Choate, "Historical Biogeography of the Pronghorn." See also Seton, *Life-Histories of Northern Animals*; McCabe, Reeves, and O'Gara, *Prairie Ghost*.

39. These dates for animal extinctions are drawn primarily from the surveys conducted by Floyd Carter for graduate thesis work at Fort Hays State University. Carter collected data from surviving settlers on the last time they had seen various species. His work was published posthumously. See Choate, "Post-Settlement History of Mammals." Some accounts put the last sighting of gray wolf and certain other mammals up to twenty-five years earlier: Cockrum, *Mammals of Kansas*. For more primary source material on settler imaginaries of wildlife in western Kansas, see the excerpts collected and commented on in Fleharty, *Wild Animals and Settlers*; see also William Hornaday, "Wild Animals in Kansas Sixty Years Ago," *Wabaunsee Truth*, February 1931.

40. Mayer and Roth, *Buffalo Harvest*, 91.

41. Mead, *Hunting and Trading*, 73–74.

42. Mayer and Roth, *Buffalo Harvest*, 32.

43. Mooar and Pace, *Buffalo Days*, 63; see also Dixon, *Life of "Billy" Dixon*, 147.

44. Isenberg, *Destruction of the Bison*, 159.

45. Mayer and Roth, *Buffalo Harvest*, 64.

46. A bill was introduced in March 1867 to tax buffalo hides but it died in committee. See Hornaday, *Extermination of the American Bison*, 519.

47. Isenberg, *Destruction of the Bison*, 132.

48. Mayer and Roth, *Buffalo Harvest*, 93.

49. For the "white lilies" comparison, see Federal Writers' Project, *Kansas*, 448. For the "charnel house" quote, see Henninger-Voss, *Animals in Human Histories*, 134.

50. Wright, *Dodge City*, 156. For more on the Kansas bone trade, see newspaper excerpts in Fleharty, *Wild Animals and Settlers*, 63–66.

51. Isenberg, *Destruction of the Bison*, 160.

52. This figure comes from a calculation by Lieutenant General Henry Inman, based on statistics from railroad freight departments. According to Inman, this represented the remains of thirty-one million bison: Inman, *Old Santa Fe Trail*, 203.

53. Cook, *Border and the Buffalo*, 102.

54. Isenberg, *Destruction of the Bison*, 160. See also Cook, *Border and the Buffalo*, 102; *Dodge City Times*, June 2, 1879, reprinted in Fleharty, *Wild Animals and Settlers*, 61–62.

55. Mayer and Roth, *Buffalo Harvest*, 92.

56. Blanchard, *Conquest of Southwest Kansas*, 202–3.

57. For general references to the madness incited by the open spaces of the Plains, see Boorstin, *Americans*, 119–20; Orpen, *Old Emigrant Days in Kansas*; Prasch, "Facing This Vast Hardness"; Webb, *Great Plains*.

58. Dodge, *Hunting Grounds of the Great West*, 47–48.

59. George W. Brown, "The Naming of Goff Creek," *Guymon Herald*, February 25, 1915, reprinted in Chrisman, *Lost Trails of the Cimarron*, 26–33.

60. Betty Wells, "Living at Larned Is a Demoralizing Experience," *Wichita Eagle*, January 28, 1974, 1.

61. Betty Wells, "Day at Larned Just One Big Trip," *Wichita Eagle*, February 2, 1974, 2A.

62. Wells's report was published between January 27 and February 2, 1974, in the *Wichita Eagle* newspaper: Betty Wells, "Inside Story of Mental Ward Told," *Wichita Eagle*, January 27, 1974; "Living at Larned Is a Demoralizing Experience," *Wichita Eagle*, January 28, 1974; "New Patient Stripped of Belongings, Dignity," *Wichita Eagle*, January 29, 1974; "Patient Welcomed by Assault, Sexual Attack," *Wichita Eagle*, January 30, 1974; "Boredom and Drugs Form a Typical Day," *Wichita Eagle*, January 31, 1974; "Hardened Aides Soon Spawn Fear in Larned Newcomer," *Wichita Eagle*, February 1, 1974; "Day at Larned Just One Big Trip," *Wichita Eagle*, February 2, 1974.

63. Kiran, Bala, and Singh, "Dental Protection," 74; Lingley and Robbins, "Fractures Following Electroshock Therapy"; Woo and Do, "Tongue Laceration."

64. The treatment was on the mind of Kansas psychiatrists and policy makers at least as early as 1950. See Group for the Advancement of Psychiatry, *Revised Electroshock Therapy Report*; Bonn and Boorstein, "Regressive Electroshock Therapy." Syndicated

medical columnist Dr. Joseph Molner, whose column appeared in several Kansas newspapers, once responded to a write-in question about electroshock by suggesting the treatment may help release people from old patterns of habit and "emotional cramps," though he admitted to not knowing exactly how and why the treatment works: Joseph Molner, "To Your Good Health: Electroshock Therapy," *Ottawa Herald*, September 26, 1959.

65. Warren, *Madwives*, 129–44.

66. Kansas Department of Agriculture, "Water Conservation Programs."

67. For more on the specifics of establishing cover on CREP land, see Conservation Reserve Enhancement Program, *Cover Crop and Grass Establishment*; Kansas Department of Agriculture, *Upper Arkansas River CREP*. For an analysis of the impediments that diminish the effectiveness of related conservation programs in southwest Kansas, see Leathers and Harrington, "Land 'Slippage.'"

68. Harrington, Lu, and Kromm, "Milking the Plains," 554.

69. Natural Resources Conservation Service, "KS652.0408 State Supplement." See also Kansas State University, *Alfalfa Production Handbook*.

70. The figure is from the turn of the millennium; the population of beef cattle is most likely significantly higher. See Harrington, Lu, and Harrington, "Fossil Water and Agriculture," 275.

71. Seaboard Foods, "Facts and Figures Processing."

72. Nilles, "Fueling the New Frontier."

73. National Beef, "Leaders in Beef Processing"; see also Tyson Foods, *Investor Fact Book*.

74. Dillon, "Clock Is Ticking." See also Kansas Leadership Center, "Largest High Plains Aquifer Users."

75. Kansas Legislative Research Department, *Kansas Corporate Farming Law*.

76. This information is compiled from figures taken from the county assessor's office, which provides irrigated and dryland acreage for each farm operator. The figure of 60% is based on a breakdown by zip code; only about 39% of land (irrigated and dryland combined) was owned by those whose primary residence was listed as Grant County. About 28% was owned by residents of other Kansas counties and about 33% was owned by out-of-state residents. This pattern of absentee or corporate landownership seems to be a much wider phenomenon across rural America. For insightful treatments of this, see Ashwood, *For-Profit Democracy*, and Ashwood et al., "What Owns the Land."

77. Worster, *Dust Bowl*, 1–63. For a cultural history of Dust Bowl eschatology, see Shumsky, "Dust, Disease, Death and Deity." For examples of first-hand observations, see Riney-Kehrberg, *Rooted in Dust*; Svobida, *Farming the Dust Bowl*. For an illuminating analysis of the Dust Bowl as a durable political, cultural, epistemic, and economic horizon that lingers into the present, see Holleman, *Dust Bowls of Empire*.

78. Kirkendall, *Social Scientists and Farm Politics*, 325.

79. A. Edwards, *Influence of Drought and Depression*.

80. Bell, *Contemporary Rural Community*.

81. A. Edwards, *Influence of Drought and Depression*, 68–75; Bell, *Contemporary Rural Community*, 1–22.

82. A. Edwards, *Influence of Drought and Depression*, 52.

83. Bell, *Contemporary Rural Community*, 106.

84. Worster, *Dust Bowl*, 1–43. See also Egan, *Worst Hard Time*.

85. Bell, *Contemporary Rural Community*, 29; A. Edwards, *Influence of Drought and Depression*, 8–9. See also Miner, *Next Year Country*.

86. A. Edwards, *Influence of Drought and Depression*, 84–94.

87. A. Edwards, 54.

88. Bell, *Contemporary Rural Community*, 107.

89. Bell, 69.

90. A. Edwards, *Influence of Drought and Depression*, 91–93.

91. Bell, *Contemporary Rural Community*, 43.

92. Bell, 75.

93. Bell, 6.

94. John A. McWethy, "Dust Bowl Defense: Farmers Ready to Fight Next Drought Cycle by Irrigation, Other Plans," *Wall Street Journal*, June 28, 1947. See also Taylor, Stewart, and Downton, "Perceptions of Drought."

95. Green, *Land of the Underground Rain*. See also Travis, "Charlatans, Sharpers, and Climatology."

96. "Valley of Death Transformed to Vale of Plenty," *Topeka Capital*, June 5, 1927.

97. "Western Kansas Eden Possible by Deep Wells," *Topeka Capital*, February 27, 1938. See also "The Rain God Defied," *Topeka Capital*, July 6, 1894.

98. John A. McWethy, "Dust Bowl Defense."

99. Mays, *Sublette Revisited*, 19.

100. Mays, 112.

101. Bell, *Contemporary Rural Community*, 75.

102. See Egan, *Worst Hard Time*.

Dust

1. The Church of Christ arose out of the American Restoration Movement, also known as the Stone-Campbell Movement. For a more detailed history, see Eckstein, *Churches of Christ in Texas*; Garrett, *Stone-Campbell Movement*; North, *Union in Truth*; R. Roberts, "Expansion of Church of Christ."

2. Much work in agricultural economics has been done on groundwater as a determinant of land values in the Ogallala region. In the context of decline, these con-

siderations are central to management decisions and models of the future. A recent study estimated that 24% of currently irrigated lands in the High Plains aquifer region may be unable to support irrigation by 2100. Of that area, 13% is projected to be not suitable for dryland agriculture owing to soil quality. See Deines et al., "Transitions from Irrigated to Dryland." See also Gilson et al., *Value of Ogallala Aquifer Water*; Golden and Johnson, "Potential Economic Impacts"; S. Lee and Bagley, "Land Values"; Torell, Libbin, and Miller, "Market Value of Water."

3. This information is contained in Southwest Kansas Groundwater Management District Number 3, "Revised (2018) (Draft) Management Program." See also Southwest Kansas Groundwater Management District Number 3, *Revised Management Program*.

4. Southwest Kansas Groundwater Management District Number 3, "Revised (2018) (Draft) Management Program," 32.

5. Southwest Kansas Groundwater Management District Number 3, 33.

6. For an overview of historical aqueduct plans and other utopian water schemes proposed as early as 1896 and as recently as 2015, see Opie, *Ogallala*, 282–317. This idea has had enduring appeal and merits greater critical attention. Importing water allows an unsustainable system of agricultural accumulation to transcend the limits of the regional ecologies to which it was historically bound. It thus promises to overcome what Sanderson and Frey, in "Desert to Breadbasket," insightfully call the "metabolic rift" at the core of aquifer depletion and its contradictory forms in southwest Kansas.

7. For general facts about WIZARD, such as data-set size and the fact that most figures are annual, see Woods and Wilson, *Water Information*. For thoughts about quantifying scale in KGS models, see Buddemeier et al., *Scale, Uncertainty*. For a discussion of the ways groundwater models may underestimate the complexity of aquifer systems, see Rojas, Feyen, and Dassargues, "Conceptual Model Uncertainty." For broader treatments of uncertainty, see Beck, *Risk Society*; Callon, Lascoumes, and Barthe, *Uncertain World*. On the implications of modeling insufficiency for anthropological thinking and writing about complex earth systems, see Petryna, "What Is a Horizon?" For applicable insights about modeling environmental crises elsewhere, see Auyero and Swistun, "Toxic Uncertainty"; Bond, "Governing Disaster"; Petryna, "Wildfires at the Edges of Science"; among others. For discussions about using satellite measurements to more accurately model depletion on the High Plains, see Longuevergne, Scanlon, and Wilson, "GRACE Hydrological Estimates"; Nie et al., "Groundwater Withdrawals"; Rodell and Famiglietti, "Satellite-Based Monitoring"; Strassberg, Scanlon, and Chambers, "Evaluation of Groundwater."

8. This information is drawn from Kansas Leadership Center, "Largest High Plains Aquifer Users." See also Dillon, "Clock Is Ticking."

9. For more on the Grant County carbon black lawsuits, see Don Kendall, "Grant Farmers Sick of Living in the Black," *Hutchinson News*, December 12, 1960. Pollution

problems continue. Additional legal actions were filed against Columbian Chemical Company in 2017 for violating emissions standards at its carbon black plants in Grant County, Kansas, and Louisiana. The complaints were filed in the US District Court for the Western District of Louisiana.

10. Newspaper reports show that the mayor of Rolla requested that state health officials investigate the record of herbicide testing in the area after five of seventeen babies born between 1975 and 1978 had birth defects. The herbicide in question was 2,4,5-T, which contained dioxin and was temporarily suspended. For more on the resulting controversy and debate, see especially Cindy Schwartz, "Did Herbicide Cause Birth Defects?: Rolla Site of a Tragic Question," *Hutchinson News*, March 28, 1979; "State Agencies Not Talking about Birth Defect Incidence," *Garden City Telegram*, March 29, 1979; "No Evidence Herbicide Was Ever Used," *Hutchinson News*, April 1, 1979; "Four Rolla Families File Birth Defect Suits," *Garden City Telegram*, May 30, 1979; "Why No Report on Rolla?" *Garden City Telegram*, June 2, 1979; "Rolla Case to Topeka," *Garden City Telegram*, March 14, 1980.

11. For an overview of research and controversy around publicizing the effects of atrazine, see Bethsass and Colangelo, "European Union Bans Atrazine." See also Hayes, "There Is No Denying This."

12. See Dirinck et al., "Obesity and Persistent Organic Pollutants"; Evangelou et al., "Exposure to Pesticides and Diabetes"; Lasserre et al., "Effects of the Endocrine Disruptors"; D.-H. Lee, "Persistent Organic Pollutants"; D.-H. Lee et al., "Low Dose Organochlorine Pesticides"; Montgomery et al., "Incident Diabetes and Pesticide Exposure"; Rodriguez, Thiruchelvam, and Cory-Slechta, "Sustained Exposure"; Slotkin, "Does Early-Life Exposure"; Swaminathan, "Pesticides and Human Diabetes"; Swanson et al., "Genetically Engineered Crops"; Velmurugan et al., "Gut Microbiota." For a groundbreaking analysis of diabetes and its relationship to planetary change, chemical thresholds, and genres of exposition, see Moran-Thomas, *Traveling with Sugar*. For a review of anthropological treatments of toxicity more broadly, see Nading, "Toxic World."

13. "Redeeming the Desert," *Commonwealth*, August 16, 1882; "The Land of Irrigation," *Lakin Index*, January 12, 1893; "Valley of Death Transformed to Vale of Plenty," *Topeka Capital*, June 5, 1927. For an illuminating environmental history of water and power in the Arkansas River valley, see Sherow, *Watering the Valley*.

14. Hyde, *Life of George Bent*, 149.

15. On the way, Chivington spoke with a freighter who had left Fort Lyon the day before. Chivington said he was after scalps. The freighter replied there were many to be had for the taking near the fort. Chivington questioned him intently about how to find these friendly Indians. Chivington, according to the freighter, said he longed to wade in gore. See Combs, "Testimony," 117.

16. [Beckwourth], "Testimony," 69.

17. Soule, "Testimony," 8–19.

18. Hyde, *Life of George Bent*, 157–58.

19. "The Battle of Sand Creek," *Rocky Mountain News*, December 17, 1864.

20. Louderback, "Testimony," 136.

21. Silas Soule, "Silas Soule to Ned Wynkoop," December 14, 1864, quoted in National Park Service, "Life of Silas Soule." More from Soule's letter: "I tell you Ned it was hard to see little children on their knees have their brains beat out by men professing to be civilized. One squaw was wounded and a fellow took a hatchet to finish her . . . and he cut one arm off, and held the other with one hand and dashed the hatchet through her brain. One squaw with her two children, were on their knees begging for their lives of a dozen soldiers, within ten feet of them all firing—when one succeeded in hitting the squaw in the thigh, then she took a knife and cut the throats of both children, and then killed herself. One old squaw hung herself in the lodge—there was not enough room for her to hang and she held up her knees and choked herself to death. Some tried to escape on the Prairie, but most of them were run down by horsemen. . . . They were all scalped, and as high as half a dozen taken from one head. They were all horribly mutilated. One woman was cut open and a child taken out of her, and scalped. White Antelope, War Bonnet and a number of others had Ears and Privates cut off. Squaw's snatches were cut out for trophies."

22. "The Fort Lyon Affair," *Rocky Mountain News*, December 30, 1864.

23. Kelman, *Misplaced Massacre*, 18.

24. Snyder, "Testimony," 77.

25. Kelman, *Misplaced Massacre*, 103.

26. Kelman, 60.

27. Kelman, 1–146.

28. See, for instance, Basso, *Wisdom Sits in Places*; Connerton, *How Modernity Forgets*; Nora and Kritzman, *Realms of Memory*; Wolfe, "Elimination of the Native." On settler innocence, see Tuck and Yang, "Decolonization." For a discussion of responsibility in a settler colonial context, see Maddison, Clark, and de Costa, *Limits of Settler Colonial Reconciliation*.

29. Chrisman, *Lost Trails of the Cimarron*, 24. See also G. Brown, *Life and Adventures*, 122.

30. Hyde, *Life of George Bent*, 46.

31. This information is drawn from Bernier, "Ethnobotany of the Northern Cheyenne," 11–12.

32. Kelman, *Misplaced Massacre*, 207.

33. Greene and Scott, *Finding Sand Creek*.

34. Whitacre and DeVore, "Cultural Resource Investigations." See also Santa Fe Trail Association, "Awards of Merit"; F. Bessire, "Wagonbed Springs."

Clouds

1. For more on overwatering, see Lilienfeld and Asmild, "Estimation of Excess Water Use."

2. This is a topic of much debate and scholarship that details technological adaptations to groundwater depletion. For a critical take on technology as the best conservation model, see Sears et al., "Jevons' Paradox." For a review of new irrigation technology being used to synch watering with crop growth cycles in depleted areas, see Himanshu et al., "Deficit Irrigation Strategies." For an overview of aquifers, economics, and agricultural adaptation in the Ogallala region, see Hornbeck and Keskin, "Historically Evolving Impact."

3. Amy Bickel, "A Fight for Water," *Hutchinson News*, September 22, 2017. For other articles that featured this important work on conservation, see Amy Bickel, "Solving the High Plains Water Crisis: With His Wells Declining, One Kansas Farmer Takes Path to Extend the Life of the Ogallala," *Hutchinson News*, September 21, 2017; Amy Bickel, "A Fight in the Field Changed One Kansas Farmer's Way of Thinking and Is Saving the Ogallala Aquifer," *High Plains Journal*, October 15, 2018; Candace Krebs, "Like Father, Like Daughter, for These Aquifer Advocates," *Ag Journal Online*, April 19, 2018, https://www.agjournalonline.com/news/20180419 /like-father-like-daughter-for-these-aquifer-advocates; Heather Clancy, "Smart Irrigation Startups."

4. Lauer and Sanderson, "Producer Attitudes towards Groundwater Conservation." For an insightful ethnography of groundwater use in western Kansas, see Gray, "Going with the Flow." For a sociological analysis of groundwater management that identifies external regulation as the least conflictive option in southwest Kansas, see Frey and Said, "Social Assessment."

5. My discussion here and throughout is informed by several analyses of the profound contradictions at the core of groundwater management in this area and beyond. It builds on prior calls for systemic shifts towards a truly sustainable socioeconomic basis. See, especially, Aistrup et al., "Conserving the Ogallala Aquifer"; Sanderson and Frey, "Desert to Breadbasket" and "Structural Impediments"; Sophocleous, "Groundwater Management Practices" and "Safe Yield." For a global treatment of groundwater management practices and fault lines, see Sagala and Smith, "Comparative Groundwater Management."

6. Northwest Kansas Groundwater Management District 4, "Request for a District-Wide LEMA," 1. See also Northwest Kansas Groundwater Management District 4, "Testimony," 2.

7. Barfield, "Order of Designation."

8. The first lawsuit, *Woofter v. McClaskey*, was dismissed by the court. Another suit, *Friesen v. Barfield*, is pending.

9. A preliminary analysis of the impacts caused by a smaller LEMA in the northwest district found that irrigated crop producers reduced groundwater use by 25.7% and changed their crops to effectively off-set the economic costs of using less water. See Golden and Liebsch, "Monitoring the Impacts."

10. This refers to grass planted as part of the Conservation Reserve Program, locally called by its abbreviation, which is a federal program that provides farmers with a yearly rental payment in exchange for removing environmentally sensitive land from agricultural production and planting ecologically appropriate grasses and legumes that will improve soil quality and habitat.

11. Gregg, *Commerce of the Prairies*, 98.

12. "Last Big Buffalo Hunt."

13. For general treatments of mental health in rural Kansas and elsewhere, see Tim Horan, "Suicide Rate High in Rural Kansas," *Salina Journal*, August 25, 2017; Houchen, Huang, and Lin, *Deaths by Suicide*; Jurich and Russell, "Family Therapy with Rural Families"; Ossian, "Bandits, Mad Men, and Suicides"; Debbie Weingarten, "Why Are America's Farmers Killing Themselves?" *Guardian*, December 11, 2018. For an overview of the connections between rural suicides and other downwardly trending life expectancies, see Deaton and Case, *Deaths of Despair*. For revelatory correlations of rural mental health with periods of environmental change and drought, see Berry et al., "Climate Change and Farmers"; B. Edwards, Gray, and Hunter, "Impact of Drought."

14. See Valdés, "Settlers, Sojourners, and Proletarians"; Oppenheimer, "Acculturation or Assimilation"; Rochín, "Latinos on the Great Plains"; M. Smith, "Beyond the Borderlands." For an analysis of the systemic inequities involved, see also S. White, "Migration Trends."

15. For an important recent ethnography of corporate hog farming, see Blanchette, *Porkopolis*. For more on racial hierarchy and industrial agriculture in western Kansas, and meat-packing in particular, see Aguilar, "Adaptation as a Process of Acquisition"; Benson, "Garden City, Kansas"; Broadway, "Meatpacking"; Champlin and Hake, "Immigration as Industrial Strategy"; Martinez, "Development of the Mexican-American Community"; Stull, "I Come to the Garden"; Stull, Broadway, and Griffith, *Any Way You Cut It*; Wahl, Gunkel, and Sanchez, "Death and Disability in the Heartland." For a discussion of the gendered hierarchies inscribed into the racialized structure of agro-industrial labor on the Plains, see Sittig and González, *Migrant Women and Meatpacking*.

16. For important accounts of the instantiation, policing, and contestation of racial boundaries in Kansas, see Campney, *This Is Not Dixie*; Correll, "Blood on the Plow"; Dickens, "Narrating Migrants and Cowboys." See also Campney, *Hostile Heartland*.

17. The bioinequities presumed within this racial and class hierarchy were laid bare during the COVID-19 pandemic, in which laborers in meat-packing plants were

compelled to continue working despite extremely high rates of infection in south-west Kansas and elsewhere. On toxic agriculture, the Environmental Protection Agency had a short-lived partnership with the Harvest America Corporation to create a program called Informing People on Pesticides, which specifically targeted seasonal laborers working in feedlots in southwestern Kansas. Though it claimed to have improved knowledge among workers while exposing potential exposures on feedlots, it appears only in the agency's 2004 biennial report on environmental justice and in no subsequent literature. The topic merits further research. See Environmental Protection Agency, *Environmental Justice Biennial Report 2004*.

18. See Marotz et al., "Cloud Fields."

19. For important treatments of this case and its implications, see Amy Bickel, "Haskell County Farmer Wins District Court Decision over Decade-Old Water Rights Battle," *Hutchinson News*, February 2, 2017; Gray, "Going with the Flow," 172–78; Griggs, "Beyond Drought" and "Political Cultures of Irrigation"; Ian James, "Kansas Water Lawsuit Could Become a Catalyst for Slowing Aquifer's Decline," *Desert Sun*, February 2, 2017; Tiffany Stecker, "Precious Commodity Sparks Fear, Feuds and Lawsuits," *E&E News*, March 25, 2016; *USA Today*, "Pumped Dry: A Race to the Bottom of the Ogallala in Kansas," YouTube, December 10, 2015, https://www.youtube.com/watch?v=7CxA8PeDhIc.

20. Here, I draw from several illuminating analyses of groundwater management frameworks in southwest Kansas. See R. Clark, "Groundwater Management"; Griggs, "Does Groundwater Management Work?"; Peck, "Groundwater Management in Kansas."

21. See case documents: Garetson Brothers v. American Warrior, Inc., 51 Kan. App. 2d 347 P.3d 687 (2015); Garetson Brothers v. American Warrior, Inc., 51 Kan. App. 2d 347 P.3d 687 (2015), *rev. denied* 303 Kan. 1077 (2016).

22. Garetson Brothers v. American Warrior, Inc., 56 Kan. App. 2d 435 P.3d 1153 (2019); see also Garetson Brothers v. American Warrior, Inc., Dist. Ct. No. 2012-CV-09 (2016).

23. For insights on the blurring of affective economies and political rationalities of "home" during moments of social upheaval, see Stoler, "'Interior Frontiers.'" For a recent example of racial violence against immigrants in southwest Kansas, see Department of Justice, "Three Southwest Kansas Men Sentenced to Prison for Plotting to Bomb Somali Immigrants in Garden City," press release, January 25, 2019.

24. See Clayton, "Weeds, People and Contested Places"; Godar and Stahlman, "Consultant's Perspective"; Kumar et al., "Confirmation of 2,4-D Resistance."

Afterword

1. Here I am paraphrasing the writings of W. G. Sebald, who wrote that he knew something was being kept from him as he came of age in the shadow of the Holo-

caust: W. G. Sebald, "A Natural History of Destruction," *New Yorker*, November 2002, 66–77. See also Sebald, *Natural History of Destruction* and *Rings of Saturn*. I draw inspiration from Sebald's techniques and insights throughout the text.

2. This is drawn from Taussig's analysis of the "hermeneutic violence" intrinsic to capitalist modes of extracting natural resources and turning them into commodities, as well as his indictment of an explanatory mode that cloaks nonsensical destruction in ideologies of civilization and moral order. See especially Taussig, *Corn Wolf*, *Shamanism*, and *Palma Africana*.

3. For illuminating analyses of groundwater management in southwest Kansas, see Griggs, "Does Groundwater Management Work?"; Peck, "Groundwater Management"; Sanderson and Frey, "Structural Impediments." For a comparative discussion of groundwater management in India, see Birkenholtz, "Groundwater Governmentality."

4. Here and throughout the text, I draw from the work of many scholars who have insightfully analyzed the general features, particular details, and wider genealogies of each of these trends. See, especially, Ashwood, *For-Profit Democracy*; Bataille, *Accursed Share*; Berger, *Dollar Harvest*; Berlage, *Farmers Helping Farmers*; Berlet and Sunshine, "Rural Rage"; Blanchette, *Porkopolis*; Champlin and Hake, "Immigration as Industrial Strategy"; Cramer, *Politics of Resentment*; Cronon, *Nature's Metropolis*; Davidson, *Broken Heartland*; Deaton and Case, *Deaths of Despair*; Edelman, "Capitalism Underdeveloped Rural America"; Fanon, *Wretched of the Earth*; Frank, *What's the Matter with Kansas?*; Gilroy, *Against Race*; Ginsburg, *Contested Lives*; Godelier, *Rationality and Irrationality*; Hochschild, *Strangers in Their Own Land*; Holleman, *Dust Bowls of Empire*; Laclau, *On Populist Reason*; Lepselter, *Resonance of Unseen Things*; Martin, *Bipolar Expeditions*; Mays, *Sublette Revisited*; McConnell, *Decline of Agrarian Democracy*; McCormick, "Secession Movement"; Moore, *Capitalism in the Web of Life*; Perramond, *Unsettled Waters*; Pisani, *From the Family Farm*; Potter, *People of Plenty*; Purdy, *This Land Is Our Land*; Roediger, *Wages of Whiteness*; Scranton, *Learning to Die*; Shapin, *Social History of Truth*; Smarsh, *Heartland*; Vogeler, *Myth of the Family Farm*; Walley, *Exit Zero*; R. White, *Organic Machine*; Worster, *Rivers of Empire*; Wuthnow, *Red State Religion*.

5. Here, I draw from James Baldwin's framing of responsibility.

6. There is a great deal of scholarship that addresses each specific case of extreme groundwater depletion from the perspective of natural science, hydrology, hydrogeology, or public policy. Here, I also draw from several important recent syntheses of groundwater loss at a global scale and its wider intersectional relationships. See especially Aeschbach-Hertig and Gleeson, "Regional Strategies"; Bierkens and Wada, "Non-renewable Groundwater Use"; Famiglietti, "Global Groundwater Crisis"; Foster and Chilton, "Groundwater"; Giordano, "Global Groundwater?"; Gleeson et al., "Global Groundwater Sustainability"; Gurdak, "Groundwater"; Konikow and Kendy, "Groundwater Depletion."

7. See, for example, Galloway and Burbey, "Regional Land Subsidence"; Minderhoud et al., "Subsidence in the Mekong Delta."

8. See especially Klassen and Allen, "Risk of Saltwater Intrusion"; Steyl and Dennis, "Coastal-Area Aquifers in Africa."

9. Among many examples, see, for instance, Debernardi et al., "Correlation between Nitrate Concentration"; Grondona et al., "Assessment of Organochlorine Pesticides"; Neh et al., "DRASTIC-GIS Model."

10. For an illuminating analysis of groundwater's status in the Green Revolution, see Moore, *Capitalism in the Web of Life*. See also Hazell, "Asian Green Revolution"; Shah et al., "Sustaining Asia's Groundwater Boom."

11. See Ibrahim, "Excessive Use of Groundwater."

12. See Van Steenbergen et al., "Groundwater Depletion in Balochistan."

13. See Changming, Jingjie, and Kendy, "Groundwater Exploitation"; Chatterjee et al., "Mapping and Management"; MacDonald et al., "Groundwater Quality and Depletion"; Rodell, Velicogna, and Famiglietti, "Satellite Based Estimates"; Shahid et al., "Groundwater Over-Exploitation"; Shankar, Kulkarni, and Krishnan, "India's Groundwater Challenge"; Shen et al., "Groundwater Depletion in the Hai River"; Shi et al., "Deep Groundwater Over-Exploitation." See also Feitelson, "Political Economy"; Komakech and de Bont, "Differentiated Access"; McKittrick, "Theories of 'Reprecipitation.'"

14. See Suter, "Yemen's Water Crisis"; see also Parker, "World's Water Wars."

15. Here and throughout, I draw from Hetherington's notion of "agri-biopolitics," which points to the relationships between monocrop cultivation and the governance of rural, agrarian populations. See Hetherington, *Government of Beans*. For more on the moral economies of biolegitimacy, see Fassin, *Humanitarian Reason*.

16. Livingston, *Self-Devouring Growth*. See also Griffiths and Robin, *Ecology and Empire*; Tsing et al., *Arts of Living*.

17. For one formulation of "groundwater citizenship," see Tiernes, "Groundwater Citizenship."

18. Here and above, I draw from a growing body of work that is re-examining various aspects of the commons in relation to planetary and social crises. See, for instance, Caffentzis and Federici, "Commons against and beyond Capitalism"; Dietz, Ostrom, and Stern, "Struggle to Govern"; Hardin, "Tragedy of the Commons"; Ostrom, *Governing the Commons*; Ostrom et al., *Drama of the Commons*; Ostrom and Gardner, "Coping with Asymmetries"; Purdy, *This Land Is Our Land*; Stern, Dietz, and Ostrom, "Research on the Commons." For important discussions of water's relationships to the notion of the commons, see Bakker, "'Commons' versus the 'Commodity'"; Ostrom, Stern, and Dietz, "Water Rights"; Provencher and Burt, "Private Property Rights Regime"; and Wagner, "Water and the Commons Imaginary."

19. Many scholars are already working in these directions. I draw inspiration from the ways many political anthropologists are using collaborative ethnography with a variety of distinct groups to better understand specific perspectives, to contest hierarchies of ascriptive difference, to challenge essentialized identities, to cross hardening social and political divides, and to build anti-authoritarian collectives opposed to injustice in its many and overlapping forms—all while more clearly placing theory at the service of those from whom it is ostensibly drawn and in the name of whose liberation it is supposedly justified.

20. Here and throughout the text, I draw from a vast range of work in anthropology that has proven this point and helped me to think more deeply and more critically about the conditions of possibility for social life on the Plains. In addition to those noted above, some especially influential accounts include Appadurai, *Fear of Small Numbers*; Biehl and Locke, *Unfinished*; Ferguson, *Give a Man a Fish*; Fischer, *Anthropology in the Meantime*; Jusionyte, *Threshold*; Li, *Land's End*; Masco, *Future of Fallout*; Petryna, *Life Exposed*; Ralph, *Torture Letters*; Roitman, *Anti-crisis*; Tsing, *Mushroom at the End of the World*.

BIBLIOGRAPHY

Adams, Gregory P., and DeRoy L. Bergman. *Geohydrology of Alluvium and Terrace Deposits of the Cimarron River from Freedom to Guthrie, Oklahoma.* Oklahoma City: US Department of the Interior, US Geological Survey, 1996.

Adler, Robert W. "Drought, Sustainability, and the Law." *Sustainability* 2, no. 7 (July 2010): 2176–96.

Aeschbach-Hertig, Werner, and Tom Gleeson. "Regional Strategies for the Accelerating Global Problem of Groundwater Depletion." *Nature Geoscience* 5 (December 2012): 853–61.

Agee, James, and Walker Evans. *Let Us Now Praise Famous Men.* Boston, MA: Houghton Mifflin, 1941.

Aguilar, Daniel. "Adaptation as a Process of Acquisition of Cultural Capital: The Case of Mexican Immigrants in Meatpacking Areas in Kansas." *Journal of Latino-Latin American Studies* 3, no. 2 (Winter 2008): 1–25.

Aistrup, Joseph A., Tom Bulatewicz, Laszlo J. Kulcsar, Jeffrey M. Peterson, Stephen M. Welch, and David R. Steward. "Conserving the Ogallala Aquifer in Southwestern Kansas: From the Wells to People, a Holistic Coupled Natural-Human Model." *Hydrology and Earth System Sciences* 21, no. 12 (2017): 6167–83.

Anand, Nikhil. *Hydraulic City: Water and the Infrastructures of Citizenship in Mumbai.* Durham, NC: Duke University Press, 2017.

Anderson, Gary Clayton. *The Conquest of Texas: Ethnic Cleansing in the Promised Land, 1820–1875.* Norman: University of Oklahoma Press, 2019.

Appadurai, Arjun. *Fear of Small Numbers: An Essay on the Geography of Anger.* Durham, NC: Duke University Press, 2006.

Ashwood, Loka. *For-Profit Democracy: Why the Government Is Losing the Trust of Rural America.* New Haven, CT: Yale University Press, 2018.

Ashwood, Loka, John Canfield, Madeleine Fairbairn, and Kathryn De Master. "What Owns the Land: The Corporate Organization of Farmland Investment." *Journal of Peasant Studies* (2020). https://doi.org/10.1080/03066150.2020.1786813.

Ashwood, Loka, Danielle Diamond, and Fiona Walker. "Property Rights and Rural Justice: A Study of US Right-to-Farm Laws." *Journal of Rural Studies* 67 (2019): 120–29.

Ashworth, William. *Ogallala Blue: Water and Life on the High Plains*. New York: W. W. Norton, 2006.

Aucoin, James. "The Irrigation Revolution and Its Environmental Consequences." *Environment* 21, no. 8 (October 1979): 17–38.

Auyero, Javier, and Debora Swistun. "The Social Production of Toxic Uncertainty." *American Sociological Review* 73 (June 2008): 357–79.

Bair, Julene. *The Ogallala Road: A Memoir of Love and Reckoning*. New York: Viking, 2014.

Bakker, Karen. "The 'Commons' versus the 'Commodity': Alter-globalization, Anti-privatization, and the Human Right to Water in the Global South." *Antipode* 39, no. 3 (June 2007): 430–55.

———. "Neoliberal versus Postneoliberal Water: Geographies of Privatization and Resistance." *Annals of the Association of American Geographers* 103, no. 2 (2013): 253–60.

———. *Privatizing Water: Governance Failure and the World's Urban Water Crisis*. Ithaca, NY: Cornell University Press, 2013.

Baldwin, James, and Margaret Mead. *A Rap on Race*. New York: Dell, 1972.

Ballestero, Andrea. *A Future History of Water*. Durham, NC: Duke University Press, 2019.

Barfield, David. "Order of Designation Regarding the Groundwater Management District No. 4 District Wide Local Enhanced Management Plan." Kansas Department of Agriculture, April 13, 2018.

———. *Rules and Regulations: Kansas Water Appropriation Act*. Topeka, KS: Kansas Department of Agriculture, 2017.

Barnes, Jessica. *Cultivating the Nile: The Everyday Politics of Water in Egypt*. Durham, NC: Duke University Press, 2014.

Basso, Keith. *Wisdom Sits in Places: Landscape and Language among the Western Apache*. Albuquerque: University of New Mexico Press, 1996.

Bataille, Georges. *The Accursed Share: An Essay on General Economy*. New York: Zone, 1991.

Beach, Sarah. "'Tractorettes' or Partners? Farmers' Views on Women in Farming Households." *Rural Sociology* 78, no. 2 (2013): 210–28.

Beach, Shawn. *Irrigation and Limited Irrigation*. Topeka, KS: US Department of Agriculture Risk Management Agency, Topeka Regional Office, 2018.

Beck, Ulrich. *Risk Society*. London: Sage, 1992.

[Beckwourth], James P. "Testimony of James P. Beckwith [*sic*]." In US War Department, *Report of the Secretary of War*, 68–76.

Bell, Earl H. *Culture of a Contemporary Rural Community: Sublette, Kansas.* Washington, DC: US Department of Agriculture, Bureau of Agricultural Economics, 1942.

Benson, Janet. "Garden City, Kansas: Vietnamese Refugees, Mexican Immigrants, and the Changing Character of a Community." In *Manifest Destinies: Americanizing Immigrants and Internationalizing Americans,* edited by David W. Haines and Carol A. Mortland, 39–54. Westport, CT: Praeger, 2001.

Berger, Samuel. *Dollar Harvest: The Story of the Farm Bureau.* Lexington, MA: Heath Lexington Books, 1971.

Berlage, Nancy. *Farmers Helping Farmers: The Rise of the Farm and Home Bureaus, 1914–1935.* Baton Rouge: Louisiana State University Press, 2016.

Berlet, Chip, and Spencer Sunshine. "Rural Rage: The Roots of Right-Wing Populism in the United States." *Journal of Peasant Studies* 46, no. 3 (2019): 480–513.

Bernier, Gabriel Ruben. "Ethnobotany of the Northern Cheyenne: Medicinal Plants." Master's thesis, University of Montana, 2004.

Berry, Helen, Anthony Hogan, Jennifer Owen, Debra Rickwood, and Lyn Fragar. "Climate Change and Farmers' Mental Health: Risks and Responses." *Asia Pacific Journal of Public Health* 23, no. 2 (2011): 119S–32S.

Bessire, Fern. *Glimpses of Grant County.* Ulysses, KS: Ulysses News, 1973.

———. "Wagonbed Springs." *Wagon Tracks* 4, no. 4 (August 1990): 15–16.

Bessire, Lucas. *Behold the Black Caiman: A Chronicle of Ayoreo Life.* Chicago: University of Chicago Press, 2014.

Bethsass, Jennifer, and Aaron Colangelo. "European Union Bans Atrazine, while the United States Negotiates Continued Use." *International Journal of Occupational and Environmental Health* 12, no. 3 (July–September 2006): 260–67.

Biehl, João, and Peter Locke, eds. *Unfinished: The Anthropology of Becoming.* Durham, NC: Duke University Press, 2017.

Bierkens, Marc, and Yoshihide Wada. "Non-renewable Groundwater Use and Groundwater Depletion: A Review." *Environmental Research Letters* 14, no. 6 (May 2019): 063002.

Birkenholtz, Trevor. "Groundwater Governmentality: Hegemony and Technologies of Resistance in Rajasthan's Groundwater Governance." *Geographical Journal* 175, no. 3 (September 2009): 208–20.

Björkman, Lisa. *Pipe Politics, Contested Waters: Embedded Infrastructures of Millennial Mumbai.* Durham, NC: Duke University Press, 2015.

Blackhawk, Ned. *Violence over the Land: Indians and Empires in the Early American West.* Cambridge, MA: Harvard University Press, 2009.

Blanchard, Leola Howard. *Conquest of Southwest Kansas: A History and Thrilling Stories of Frontier Life in the State of Kansas.* Wichita, KS: Wichita Eagle, 1931.

Blanchette, Alex. *Porkopolis: American Animality, Standardized Life, and the Factory Farm.* Durham, NC: Duke University Press, 2020.

Bond, David. "Governing Disaster: The Political Life of the Environment during the BP Oil Spill." *Cultural Anthropology* 28, no 4 (2014): 694–715.

Bonn, Ethel, and Seymour Boorstein. "Regressive Electroshock Therapy and Anaclitic Psychotherapy." *Bulletin of the Menninger Clinic* 23, no. 5 (September 1, 1959): 190–201.

Boorstin, Daniel J. *The Americans: The Democratic Experience.* New York: Vintage Books, 1974.

Brauer, David, Dan Devlin, Kevin Wagner, Mike Ballou, Dean Hawkins, and Robert Lascano. "Ogallala Aquifer Program: A Catalyst for Research and Education to Sustain the Ogallala Aquifer on the Southern High Plains (2003–2017)." *Journal of Contemporary Water Research and Education* 162, no. 1 (December 2017): 4–17.

Bredehoeft, John D., Stephen S. Papadopulos, and H. H. Cooper Jr. "Groundwater: The Water-Budget Myth." In *Scientific Basis of Water Resource Management,* by National Research Council, 51–57. Washington, DC: National Academies Press, 1982.

Broadway, Michael. "Meatpacking and the Transformation of Rural Communities: A Comparison of Brooks, Alberta and Garden City, Kansas." *Rural Sociology* 72, no. 4 (December 2007): 560–82.

Brown, George W. *Life and Adventures of George W. Brown: Soldier, Pioneer, Scout, Plainsman and Buffalo Hunter.* Edited by William Elsey Connelley. Topeka, KS: Kansas State Historical Society, 1928.

Brown, J. Christopher, Eric Hanley, Jason Bergtold, Marcelus Caldas, Vijay Barve, Dana Peterson, Ryan Callihan, et al. "Ethanol Plant Location and Intensification vs. Extensification of Corn Cropping in Kansas." *Applied Geography* 53 (September 1, 2014): 141–48.

Buchanan, Rex C., Robert W. Buddemeier, and B. Brownie Wilson. *The High Plains Aquifer.* Kansas Geological Survey Public Information Circular. Lawrence, KS: Kansas Geological Survey, February 19, 2015.

Buddemeier, R. W., B. B. Wilson, J. Mosteller, and G. R. Hecox. *Scale, Uncertainty, and the Relationships between Basic Data, Information, and Management Perspectives.* KGS Open File Report, Technical Series 25. Lawrence, KS: Kansas Geological Survey, 2002.

Butler, James, Jr., Donald O. Wittemore, B. Brownie Wilson, and Geoffrey C. Bohling. "Sustainability of Aquifers Supporting Irrigated Agriculture: A Case Study of the High Plains Aquifer in Kansas." *Water International* 43, no. 6 (2018): 815–28.

Butler, William Francis. *Sir William Butler: An Autobiography.* New York: Charles Scribner's Sons, 1913.

Caffentzis, George, and Silvia Federici. "Commons against and beyond Capitalism." *Community Development Journal* 49, no. 1 (January 2014): 92–105.

Callon, Michel, Pierre Lascoumes, and Yannick Barthe. *Acting in an Uncertain World: An Essay on Technical Democracy.* Translated by Graham Burchell. Cambridge, MA: MIT Press, 2009.

Campney, Brent M. S. *Hostile Heartland: Racism, Repression and Resistance in the Midwest.* Urbana: University of Illinois Press, 2019.

———. *This Is Not Dixie: Racist Violence in Kansas, 1861–1927.* Urbana: University of Illinois Press, 2015.

Chakrabarty, Dipesh. "The Climate of History: Four Theses." *Critical Inquiry* 35, no. 2 (2009): 197–222.

Chalfant, William Y. *Cheyennes at Dark Water Creek: The Last Fight of the Red River War.* Norman: University of Oklahoma Press, 1997.

Champlin, Dell, and Eric Hake. "Immigration as Industrial Strategy in American Meatpacking." *Review of Political Economy* 18, no. 1 (2006): 49–70.

Changming, Liu, Yu Jingjie, and Eloise Kendy. "Groundwater Exploitation and Its Impact on the North China Plain." *Water International* 26, no. 2 (2001): 265–72.

Chatterjee, Rana, A. K. Jain, Subhas Chandra, Vikas Tomar, P. K. Parchure, and Shakeel Ahmed. "Mapping and Management of Aquifers Suffering from Over-Exploitation of Groundwater Resources in Baswa-Bandikui Watershed, Rajasthan, India." *Environmental Earth Sciences* 77, no. 5 (March 2018): 1–14.

Choate, Jerry R. "Post-Settlement History of Mammals in Western Kansas." *Southwestern Naturalist* 32, no. 2 (June 15, 1987): 157–68.

Chrisman, Harry E. *Lost Trails of the Cimarron.* Denver, CO: Sage Books, 1961.

Ciampitti, Ignacio A., Daniel O'Brien, Curtis Thompson, Danny Rogers, Ajay Sharda, Sarah Zukoff, Dorivar Ruiz Dias, Doug Jardine, and Brian McCornack. *Kansas Corn Management 2019.* Manhattan, KS: Kansas State University Agricultural Experiment Station and Cooperative Extension Service, December 2018.

Claassen, Roger, Christian Langpap, and JunJie Wu. "Impacts of Federal Crop Insurance on Land Use and Environmental Quality." *American Journal of Agricultural Economics* 99, no. 3 (2017): 592–613.

Clancy, Heather. "Why Smart Irrigation Startups Are Bubbling Up." *GreenBiz Group* (blog), March 22, 2018. https://www.greenbiz.com/article/why-smart-irrigation -startups-are-bubbling.

Clark, Robert Emmet. "Groundwater Management: Law and Local Response." *Arizona Law Review* 6 (1964): 178–211.

Clark, William P. *The Indian Sign Language.* Philadelphia: L. R. Hammersly, 1885.

Clayton, Neil. "Weeds, People and Contested Places." *Environment and History* 9, no. 3 (August 2003): 301–31.

Cockrum, E. L. *Mammals of Kansas.* Lawrence: University of Kansas Publications, Museum of Natural History, 1952.

Combs, James M. "Testimony of James M. Combs." In US War Department, *Report of the Secretary of War*, 115–34.

Condon, Nicole, Heather Klemick, and Ann Wolverton. "Impacts of Ethanol Policy on Corn Prices: A Review and Meta-analysis of Recent Evidence." *Food Policy* 51 (February 1, 2015): 63–73.

Connerton, Paul. *How Modernity Forgets*. Cambridge: Cambridge University Press, 2009.

Conservation Reserve Enhancement Program. *Guidelines for Cover Crop and Grass Establishment on Sandy Sites*. CRP Technical Guidance Document 84. Natural Resources Conservation Service Kansas, 2013. https://www.nrcs.usda.gov /Internet/FSE_DOCUMENTS/nrcs142p2_030412.pdf.

Cook, John R. *The Border and the Buffalo, an Untold Story of the Southwest Plains*. Topeka, KS: Crane, 1907.

Correll, Caleb. "Blood on the Plow: Extremist Group Activity during the 1980s Farm Crisis in Kansas." PhD dissertation, University of Kansas, 2019.

Cozzens, Peter. *The Earth Is Weeping: The Epic Story of the Indian Wars for the American West*. New York: Vintage Books, 2017.

Cramer, Katherine J. *The Politics of Resentment: Rural Consciousness in Wisconsin and the Rise of Scott Walker*. Chicago: University of Chicago Press, 2016.

Cronon, William. *Nature's Metropolis: Chicago and the Great West*. New York: W. W. Norton, 2009.

Cruse, Richard M., Daniel L. Devlin, Doug Parker, and Reagan M. Waskom. "Irrigation Aquifer Depletion: The Nexus Linchpin." *Journal of Environmental Studies and Sciences* 6, no. 1 (March 1, 2016): 149–60.

Darton, Nelson H. *Preliminary Report on the Geology and Underground Water Resources of the Central Great Plains*. United States Geological Society Professional Paper 32. Washington, DC: Government Printing Office, 1905.

———. *Preliminary Report on the Geology and Water Resources of Nebraska West of the One Hundred and Third Meridian*. United States Geological Society Professional Paper 17. Washington, DC: Government Printing Office, 1903.

Davidson, Osha. *Broken Heartland: The Rise of America's Rural Ghetto*. New York: Anchor Books, 1990.

Deaton, Angus, and Anne Case. *Deaths of Despair and the Future of Capitalism*. Princeton, NJ: Princeton University Press, 2020.

Debernardi, Laura, Domenico Antonio De Luca, and Manuela Lasagna. "Correlation between Nitrate Concentration in Groundwater and Parameters Affecting Aquifer Intrinsic Vulnerability." *Environmental Geology* 55, no. 3 (September 2008): 539–58.

Deines, Jillian, Meagan E. Schipanski, Bill Golden, Samuel C. Zipper, Soheil Nozari, Caitlin Rottler, Bridget Guerrero, and Vaishali Sharda. "Transitions from Irrigated to Dryland Agriculture in the Ogallala Aquifer: Land Use Suitability and Regional Economic Impacts." *Agricultural Water Management* 233 (April 2020): 106061.

Deryugina, Tatyana, and Megan Konar. "Impacts of Crop Insurance on Water Withdrawals for Irrigation." *Advances in Water Resources* 110 (December 1, 2017): 437–44.

De Wit, Cary W. "Women's Sense of Place on the American High Plains." *Great Plains Quarterly* 21, no. 1 (Winter 2001): 29–44.

Dickens, Addison R. "Narrating Migrants and Cowboys: Language and Race in Southwest Kansas Local Government." PhD dissertation, University of California–Los Angeles, 2019.

Dietz, Thomas, Elinor Ostrom, and Paul C. Stern. "The Struggle to Govern the Commons." *Science* 302, no. 5652 (2003): 1907–12.

Dillon, Karen. "The Clock Is Ticking on Dealing with Western Kansas' Drying Ogallala Aquifer." *Kansas Leadership Center Journal*, August 10, 2018. https://klcjournal.com /the-clock-is-ticking-on-dealing-with-western-kansas-drying-ogallala-aquifer/.

Dirinck, Eveline, Philippe G. Jorens, Adrian Covaci, Tinne Geens, Laurence Roosens, Hugo Neels, Ilse Mertens, and Luc Van Gaal. "Obesity and Persistent Organic Pollutants: Possible Obesogenic Effect of Organochlorine Pesticides and Polychlorinated Biphenyls." *Obesity* 19, no. 4 (April 2011): 709–14.

Dixon, Olive K. *Life of "Billy" Dixon, Plainsman, Scout and Pioneer.* Dallas: P. L. Turner, 1927.

Dodge, Richard I. *The Hunting Grounds of the Great West: A Description of the Plains, Game, and Indians of the Great North American Desert.* London: Chatto and Windus, 1877.

Domina, David A. "Federal Crop Insurance: What It Means and How It Works." *Nebraska Lawyer* 15 (November/December 2012): 11–18.

Drabik, Dušan, Pavel Ciaian, and Ján Pokrivcák. "The Effect of Ethanol Policies on the Vertical Price Transmission in Corn and Food Markets." *Energy Economics* 55 (March 1, 2016): 189–99.

Eckstein, Stephen Daniel. *History of the Churches of Christ in Texas, 1824–1950.* Austin, TX: Firm Foundation, 1963.

Edelman, Marc. "How Capitalism Underdeveloped Rural America." *Jacobin*, January 26, 2020. https://jacobinmag.com/2020/01/capitalism-underdeveloped -rural-america-trump-white-working-class.

Edwards, Allen D. *Influence of Drought and Depression on a Rural Community: A Case Study in Haskell County, Kansas.* Washington, DC: US Department of Agriculture, Farm Security Administration, 1939.

Edwards, Ben, Matthew Gray, and Boyd Hunter. "The Impact of Drought on Mental Health in Rural and Regional Australia." *Social Indicators Research* 121, no. 1 (2015): 177–94.

Egan, Timothy. *The Worst Hard Time: The Untold Story of Those Who Survived the Great American Dust Bowl*. New York: Houghton Mifflin Harcourt, 2006.

Eklund, Elizabeth Marie. "Where the Water Turns: Water and Cultural Memory in the Sierra Madre Foothills." PhD dissertation, University of Arizona, 2020.

Environmental Protection Agency. *Environmental Justice Biennial Report 2004: Integrating Environmental Justice*. Washington, DC: Environmental Protection Agency, 2004.

Espy, James P. *The Philosophy of Storms*. Boston: C. C. Little and J. Brown, 1841.

Evangelou, Evangelos, Georgios Ntritsos, Maria Chondrogiorgi, Fotini K. Kavvoura, Antonio F. Hernandez, Evangelia E. Ntzani, and Ioanna Tzoulaki. "Exposure to Pesticides and Diabetes: A Systematic Review and Meta-analysis." *Environment International* 91 (May 2016): 60–68.

Fabricant, Nicole, and Kathryn Hicks. "Bolivia's Next Water War: Historicizing the Struggles over Access to Water Resources in the Twenty-First Century." *Radical History Review* 2013, no. 116 (2013): 130–45.

Famiglietti, James S. "The Global Groundwater Crisis." *Nature Climate Change* 4, no. 11 (November 2014): 945–48.

Famiglietti, James S., Minhui Lo, Sing L. Ho, James Bethune, K. H. Anderson, Tajdarul H. Syed, Sean C. Swenson, Caroline R. de Linage, and Matthew Rodell. "Satellites Measure Recent Rates of Groundwater Depletion in California's Central Valley." *Geophysical Research Letters* 38, no. 3 (2011): L03403.

Fanon, Frantz. *The Wretched of the Earth*. New York: Grove, 1963.

Fassin, Didier. *Critique of Humanitarian Reason*. Berkeley: University of California Press, 2012.

Federal Writers' Project of the Work Projects Administration for the State of Kansas. *Kansas: A Guide to the Sunflower State*. New York: Hastings House, 1949.

Feitelson, Eran. "Political Economy of Groundwater Exploitation: The Israeli Case." *International Journal of Water Resources Development* 21, no. 3 (2005): 413–23.

Feng, Wei, Min Zhong, Jean-Michel Lemoine, Richard Biancale, Hou-Tse Hsu, and Jun Xia. "Evaluation of Groundwater Depletion in North China Using the Gravity Recovery and Climate Experiment (GRACE) Data and Ground-Based Measurements." *Water Resources Research* 49, no. 4 (April 2013): 2110–18.

Ferguson, James. *Give a Man a Fish: Reflections on the New Politics of Distribution*. Durham, NC: Duke University Press, 2015.

Fischer, Michael. *Anthropology in the Meantime*. Durham, NC: Duke University Press, 2018.

Fleharty, Eugene. *Wild Animals and Settlers on the Great Plains*. Norman: University of Oklahoma Press, 1995.

Flora, Cornelia Butler, and Jan L. Flora. "Structure of Agriculture and Women's Culture in the Great Plains." *Great Plains Quarterly* 8, no. 4 (Fall 1988): 195–205.

Flores, Dan. "Bison Ecology and Bison Diplomacy: The Southern Plains from 1800 to 1850." *Journal of American History* 78, no. 2 (September 1991): 465–85.

Foster, S., and P. J. Chilton. "Groundwater: The Processes and Global Significance of Aquifer Degradation." *Philosophical Transactions of the Royal Society of London B Biological Sciences* 358, no. 1440 (December 29, 2003): 1957–72.

Fouquet, L. C. "Buffalo Days." *Collections of the Kansas State Historical Society* 16 (1925): 341–52.

Frank, Thomas. *What's the Matter with Kansas? How Conservatives Won the Heart of America*. New York: Metropolitan Books, 2004.

Frey, R. Scott, and Ismail Said. "A Social Assessment of Alternative Water Management Policies in Southwestern Kansas." *Journal of Environmental Systems* 18, no. 2 (1988–89): 173–82.

Galchen, Rivka. "Weather Underground." *New Yorker* 13 (2015): 34–40.

Galloway, Devin, and Thomas Burbey. "Regional Land Subsidence Accompanying Groundwater Extraction." *Hydrogeology Journal* 19, no. 8 (December 2011): 1459–86.

Gard, Wayne. *The Great Buffalo Hunt*. New York: Knopf, 1959.

———. "The Mooar Brothers, Buffalo Hunters." *Southwestern Historical Quarterly* 63, no. 1 (July 1959): 31–45.

Garrett, Leroy. *The Stone-Campbell Movement: The Story of the American Restoration Movement*. Joplin, MO: College Press, 1994.

Gibson, Jane, and Benjamin Gray. "Regulating the Ogallala: Paradox and Ambiguity in Western Kansas." *Research in Economic Anthropology* 36 (2016): 3–32.

Gilpin, William. *The Central Gold Region*. Philadelphia: J. B. Lippincott, 1873.

Gilroy, Paul. *Against Race*. Cambridge, MA: Harvard University Press, 2000.

Gilson, Preston, Joseph A. Aistrup, John Heinrichs, and Brett Zollinger. *The Value of Ogallala Aquifer Water in Southwest Kansas*. Docking Institute of Public Affairs. Hays, KS: Fort Hays State University, 2001.

Ginsburg, Faye. *Contested Lives: The Abortion Debate in an American Community*. Berkeley: University of California Press, 1989.

Giordano, Mark. "Global Groundwater? Issues and Solutions." *Annual Review of Environment and Resources* 34 (2009): 153–78.

Gleeson, Tom, Mark Cuthbert, Grant Ferguson, and Debra Perrone. "Global Groundwater Sustainability, Resources, and Systems in the Anthropocene." *Annual Review of Earth and Planetary Sciences* 48 (May 2020): 431–63.

Gleeson, Tom, Yoshihide Wada, Marc Bierkens, and Ludovicus P. H. Van Beek. "Water Balance of Global Aquifers Revealed by Groundwater Footprint." *Nature* 488, no. 7410 (August 8, 2012): 197–200.

Gleick, Peter H. "The Human Right to Water." *Water Policy* 1, no. 5 (1998): 487–503.

Godar, Amar S., and Phillip W. Stahlman. "Consultant's Perspective on the Evolution and Management of Glyphosate-Resistant Kochia (*Kochia scoparia*) in Western Kansas." *Weed Technology* 29, no. 2 (June 2015): 318–28.

Godelier, Maurice. *Rationality and Irrationality in Economics.* London: New Left Books, 1972.

Golden, Bill, and Jeff Johnson. "Potential Economic Impacts of Water-Use Changes in Southwest Kansas." *Journal of Natural Resources Policy Research* 5, no. 2–3 (2013): 129–45.

Golden, Bill, and Kelln Liebsch. "Monitoring the Impacts of Sheridan County 6 Local Enhanced Management Area: Interim Report for 2013–2016." In *Proceedings of the 30th Annual Central Plains Irrigation Conference*, 38–55. Colby, KS: Central Plains Irrigation Association, 2018.

Goldman, Francisco. *The Art of Political Murder: Who Killed the Bishop?* New York: Grove/Atlantic, 2008.

Gordillo, Gastón. *Landscapes of Devils: Tensions of Place and Memory in the Argentinian Chaco.* Durham, NC: Duke University Press, 2006.

———. *Rubble: The Afterlife of Destruction.* Durham, NC: Duke University Press, 2014.

Grant, Michael Johnston. *Down and Out on the Family Farm: Rural Rehabilitation in the Great Plains, 1929–1945.* Lincoln: University of Nebraska Press, 2002.

Gray, Benjamin J. "Going with the Flow: Southwest Kansas Farmers and the Declining Ogallala Aquifer." PhD dissertation, University of Kansas, 2016.

Green, Donald. *Land of the Underground Rain: Irrigation on the Texas High Plains, 1910–1970.* Austin: University of Texas Press, 1973.

Greene, Jerome A., and Douglas D. Scott. *Finding Sand Creek: History, Archeology, and the 1864 Massacre Site.* Norman: University of Oklahoma Press, 2004.

Gregg, Josiah. *The Commerce of the Prairies: Journal of a Santa Fe Trader.* Vol. 1. New York: Henry Langley, 1844.

Griffiths, Tom, and Libby Robin, eds. *Ecology and Empire: Environmental History of Settler Societies.* Seattle: University of Washington Press, 1997.

Griggs, Burke. "Beyond Drought: Water Rights in the Age of Permanent Depletion." *University of Kansas Law Review* 62 (June 1, 2014): 1263–324.

———. "Does Groundwater Management Work?" *Kansas Journal of Law and Public Policy* 15 (2005): 391–406.

———. "The Political Cultures of Irrigation and the Proxy Battles of Interstate Water Litigation." *Natural Resources Journal* 57, no. 1 (Winter 2017): 1–74.

Grinnell, George Bird. "Cheyenne Stream Names." *American Anthropologist* 8, no. 1 (1906): 15–22.

Grondona, Sebastián, Mariana Gonzalez, Daniel Emilio Martinez, Hector Enrique Massone, and Karina Silvia Beatriz Miglioranza. "Assessment of Organochlorine

Pesticides in Phreatic Aquifer of Pampean Region, Argentina." *Bulletin of Environmental Contamination and Toxicology* 102, no. 4 (April 2019): 544–49.

Group for the Advancement of Psychiatry, Committee on Therapy. *Revised Electroshock Therapy Report No. 15.* Topeka, KS: Group for the Advancement of Psychiatry, Committee on Therapy, August 1950.

Gurdak, Jason. "Groundwater: Climate-Induced Pumping." *Nature Geoscience* 10, no. 2 (February 2017): 71–72.

Gutentag, Edwin, Fredrick J. Heimes, Noel C. Krothe, Richard R. Luckey, and John B. Weeks. "Geohydrology of the High Plains Aquifer in Parts of Colorado, Kansas, Nebraska, New Mexico, Oklahoma, South Dakota, Texas and Wyoming." US Geological Survey Professional Paper 1400-B. Washington, DC: Government Printing Office, 1984.

Hämäläinen, Pekka. *The Comanche Empire.* New Haven, CT: Yale University Press, 2008.

Hanner, John. "Government Response to the Buffalo Hide Trade, 1871–1883." *Journal of Law and Economics* 24, no. 2 (October 1981): 239–71.

Hardin, Garrett. "The Tragedy of the Commons." *Science* 162, no. 3859 (December 13, 1968): 1243–48.

Harrington, Lisa, Max Lu, and John A. Harrington Jr. "Fossil Water and Agriculture in Southwestern Kansas." In *Sustainable Communities on a Sustainable Planet: The Human-Environment Regional Observatory Project*, edited by Colin Polsky, James O'Brien, and Brent Yarnal, 270–92. Cambridge: Cambridge University Press, 2009.

Harrington, Lisa, Max Lu, and David E. Kromm. "Milking the Plains: Movement of Large Dairy Operations into Southwestern Kansas." *Geographical Review* 100, no. 4 (October 2010): 538–58.

Hastrup, Frida, and Kirsten Hastrup, eds. *Waterworlds: Anthropology in Fluid Environments.* New York: Berghahn Books, 2016.

Hayes, Tyrone. "There Is No Denying This: Defusing the Confusion about Atrazine." *Bioscience* 54, no. 12 (December 2004): 1138–49.

Hazell, Peter. "The Asian Green Revolution." International Food Policy Research Institute Discussion Paper Number 00911. Washington, DC, 2009.

Helmreich, Stefan. "Waves: An Anthropology of Scientific Things." *HAU: Journal of Ethnographic Theory* 4, no. 3 (2014): 265–84.

Henninger-Voss, Mary J. *Animals in Human Histories: The Mirror of Nature and Culture.* Rochester, NY: University of Rochester Press, 2002.

Hetherington, Kregg. *The Government of Beans: Regulating Life in the Age of Monocrops.* Durham, NC: Duke University Press, 2020.

Himanshu, Sushil Kumar, Srinivasulu Ale, James Bordovsky, and Murali Darapuneni. "Evaluation of Crop-Growth-Stage-Based Deficit Irrigation Strategies for Cotton

Production in the Southern High Plains." *Agricultural Water Management* 225 (November 2019): 105782.

Hochschild, Arlie. *Strangers in Their Own Land*. New York: New Press, 2016.

Holleman, Hannah. *Dust Bowls of Empire: Imperialism, Environmental Politics, and the Injustice of "Green" Capitalism*. New Haven, CT: Yale University Press, 2018.

Hornaday, William T. *The Extermination of the American Bison*. Washington, DC: Government Printing Office, 1889.

Hornbeck, Richard, and Pinar Keskin. "The Historically Evolving Impact of the Ogallala Aquifer: Agricultural Adaptation to Groundwater and Drought." *American Economic Journal: Applied Economics* 6, no. 1 (January 2014): 190–219.

Houchen, Carlie, Cheng-Chung Huang, and Wen-Chieh Lin. *Deaths by Suicide: Disparities in the Sunflower State*. Issue Briefs. Topeka, KS: Kansas Health Institute, April 2018.

Huang, Zhiyong, Yun Pan, Huili Gong, Pat J. F. Yeh, Xiaojuan Li, Demin Zhou, and Wenji Zhao. "Subregional-Scale Groundwater Depletion Detected by GRACE for Both Shallow and Deep Aquifers in North China Plain." *Geophysical Research Letters* 42, no. 6 (March 2015): 1791–99.

Hunt, Will. *Underground: A Human History of the Worlds Beneath Our Feet*. New York: Random House, 2020.

Hyde, George. *Life of George Bent: Written from His Letters*. Norman: University of Oklahoma Press, 1968.

Hyman, Ray, and Elizabeth G. Cohen. "Water-Witching in the United States." *American Sociological Review* 22, no. 6 (December 1957): 719–24.

Hyndman, Donald, and David Hyndman. *Natural Hazards and Disasters*. Belmont, CA: Brooks/Cole, 2014.

Ibrahim, Abdulla Ali al-. "Excessive Use of Groundwater Resources in Saudi Arabia: Impacts and Policy Options." *Ambio* 20, no. 1 (February 1991): 34–37.

Inman, Henry. *The Old Santa Fe Trail: The Story of a Great Highway*. Topeka, KS: Crane, 1916.

Isenberg, Andrew C. *The Destruction of the Bison: An Environmental History, 1750–1920*. Cambridge: Cambridge University Press, 2000.

James, Ian, and Steve Elfers. "Pumped beyond Limits, Many U.S. Aquifers in Decline." *USA Today*, December 10, 2015. https://www.usatoday.com/pages/interactives/groundwater/.

Johnston, Barbara Rose. "The Political Ecology of Water: An Introduction." *Capitalism Nature Socialism* 14, no. 3 (September 2003): 73–90.

Jurich, Anthony, and Candyce S. Russell. "Family Therapy with Rural Families in a Time of Farm Crisis." *Family Relations* 36, no. 4 (October 1987): 364–67.

Jusionyte, Ieva. *Threshold: Emergency Responders on the US-Mexico Border*. Berkeley: University of California Press, 2018.

Kane, Stephanie. *Where Rivers Meet the Sea: The Political Ecology of Water*. Philadelphia, PA: Temple University Press, 2012.

Kansas Corn. "Building Markets for Ethanol." Accessed October 17, 2020. https://kscorn.com/ethanol/.

Kansas Department of Agriculture. *Upper Arkansas River CREP Performance Report*. Manhattan, KS: Kansas Department of Agriculture, Division of Conservation, 2017.

Kansas Department of Agriculture, Division of Conservation. "Water Conservation Programs." February 21, 2019; last updated June 3, 2020. https://agriculture.ks.gov/divisions-programs/division-of-conservation/water-conservation-programs.

Kansas Department of Agriculture, Division of Water Resources. *Final Report of the Fact Finder/Referee Pertaining to Case No. 12-CV-9 District Court of Haskell County*. Topeka, KS: Kansas Department of Agriculture, Division of Water Resources, March 27, 2014.

———. "Southwest Kansas GMD Number 3." Accessed May 3, 2019. https://agriculture.ks.gov/divisions-programs/dwr/managing-kansas-water-resources/groundwater-management-districts/gmd-no.-3.

Kansas Geological Survey. "High Plains Aquifer Interactive Atlas." Accessed September 13, 2019. http//www.kgs.ku.edu/HighPlains/HPA_Atlas/InteractiveAtlas.html.

Kansas Grain Sorghum Commission. "About Ethanol." Accessed October 21, 2020. https://ksgrainsorghum.org/ethanol/.

Kansas Leadership Center. "The 150 Largest High Plains Aquifer Users in Kansas." *Kansas Leadership Center Journal*, August 10, 2018. https://klcjournal.com/the-150-largest-high-plains-aquifer-users-in-kansas/.

Kansas Legislative Research Department. *Kansas Corporate Farming Law*. Agriculture and Natural Resources Publications. Topeka, KS: Kansas Legislative Research Department, 2013.

Kansas State Board of Agriculture. *Twenty-Third Biennial Report of the Kansas State Board of Agriculture to the Legislature of the State for the Years 1921 and 1922*. Topeka, KS: Kansas State Board of Agriculture, 1922.

Kansas State University Agricultural Experiment Station and Cooperative Extension Service. *Alfalfa Production Handbook*. C683 (revised). Manhattan, KS: Kansas State University Agricultural Experiment Station and Cooperative Extension Service, 1998.

———. *Corn Production Handbook*. C560 (revised). Manhattan, KS: Kansas State University Agricultural Experiment Station and Cooperative Extension Service, 1994.

———. *Corn Production Handbook*. C560 (revised). Manhattan, KS: Kansas State University Agricultural Experiment Station and Cooperative Extension Service, 2007.

Kansas Water Office. "USDA Crop Insurance—Limited Irrigation." Accessed February 24, 2019. www.kwo.ks.gov/projects/usda-crop-insurance---limited-irrigaton.

Kauffman, Nathan, and Ty Kreitman. "Large-Scale Financing Drives Ag Lending Activity Higher." *Main Street Views: Policy Insights from the Kansas City Fed* (blog), October 19, 2018. www.kansascityfed.org/research/indicatorsdata /agfinancedatabook/articles/2018/10-17-2018/ag-finance-dbk-10-19-2018.

Kelly, William. "Concepts in the Anthropological Study of Irrigation." *American Anthropologist* 85, no. 4 (December 1983): 880–86.

Kelman, Ari. *A Misplaced Massacre: Struggling over the Memory of Sand Creek.* Cambridge, MA: Harvard University Press, 2013.

Kepfield, Sam S. "The 'Liquid Gold' Rush: Groundwater Irrigation and Law in Nebraska, 1900–93." *Great Plains Quarterly* 13, no. 4 (Fall 1993): 237–49.

Kilian, Edward. "Observations of the Divining Rod." In *Transactions of the Twentieth and Twenty-First Annual Meetings of the Kansas Academy of Science, 1887–88,* by Kansas Academy of Science, 114–15. Topeka, KS: Kansas Publishing House, 1889.

Kiran, Shashi, Renu Bala, and Tarandeep Singh. "Dental Protection during Modified Electroconvulsive Therapy Using Roll-Gauze Mouth Gag." *Journal of ECT* 25, no. 1 (March 2009): 74–75.

Kirkendall, Richard S. *Social Scientists and Farm Politics in the Age of Roosevelt.* Columbia: University of Missouri Press, 1966.

Kisekka, I., A. Schlegel, Liwang Ma, P. H. Gowda, and P.V.V. Prasad. "Optimizing Preplant Irrigation for Maize under Limited Water in the High Plains." *Agricultural Water Management* 187 (June 1, 2017): 154–63.

Klassen, J., and D. Allen. "Assessing the Risk of Saltwater Intrusion in Coastal Aquifers." *Journal of Hydrology* 551 (August 2017): 730–45.

Komakech, Hans C., and Chris de Bont. "Differentiated Access: Challenges of Equitable and Sustainable Groundwater Exploitation in Tanzania." *Water Alternatives* 11, no. 3 (2018): 623–37.

Konikow, Leonard, and Eloise Kendy. "Groundwater Depletion: A Global Problem." *Hydrogeology Journal* 13, no. 1 (March 2005): 317–20.

Kramer, Jane. *The Last Cowboy.* New York: Random House, 1975.

Krautwurst, Udo. "Water-Witching Modernist Epistemologies and Dowsing Rationality: Exporting Models of Non-rationality through Colonial and Development Discourses." *PoLAR: Political and Legal Anthropology Review* 21, no. 2 (November 1, 1998): 71–82.

Kromm, David E., and Stephen E. White, eds. *Groundwater Exploitation in the High Plains.* Lawrence: University Press of Kansas, 1992.

Kumar, Vipan, Rui Liu, Grant Boyer, and Phillip W. Stahlman. "Confirmation of 2,4-D Resistance and Identification of Multiple Resistance in a Kansas Palmer

Amaranth (*Amaranthus palmeri*) Population." *Pest Management Science* 75, no. 11 (November 2019): 2925–33.

Laclau, Ernesto. *On Populist Reason*. London: Verso, 2005.

Lasserre, Jean-Paul, Fred Fack, Dominique Revets, Sebastien Planchon, Jenny Renaut, Lucien Hoffmann, Arno C. Gutleb, Claude P. Muller, and Torsten Bohn. "Effects of the Endocrine Disruptors Atrazine and PCB 153 on the Protein Expression of MCF-7 Human Cells." *Journal of Proteome Research* 8, no. 12 (December 2009): 5485–96.

"The Last Big Buffalo Hunt in the Llano Estacado." 1887, University of Oklahoma Libraries, Western History Collections, Walter Stanley Campbell Collection.

Lauer, Stephen, and Matthew Sanderson. "Producer Attitudes towards Groundwater Conservation in the US Ogallala-High Plains." *Groundwater* 58, no 4 (July 2020): 674–80.

Leathers, Nancy, and Lisa M. B. Harrington. "Effectiveness of Conservation Reserve Programs and Land 'Slippage' in Southwestern Kansas." *Professional Geographer* 52, no. 1 (2000): 83–93.

Lee, Duk-Hee. "Persistent Organic Pollutants and Obesity-Related Metabolic Dysfunction: Focusing on Type 2 Diabetes." *Epidemiology and Health* 34 (2012): e2012002.

Lee, Duk-Hee, Michael W. Steffes, Andreas Sjödin, Richard S. Jones, Larry L. Needham, and David R. Jacobs Jr. "Low Dose Organochlorine Pesticides and Polychlorinated Biphenyls Predict Obesity, Dyslipidemia, and Insulin Resistance among People Free of Diabetes." *PLOS ONE* 6, no. 1 (2011): e15977.

Lee, Samuel H., and Edgar S. Bagley. "Groundwater and Land Values in Southwestern Kansas." *Ground Water* 10, no. 6 (November–December 1972): 27–36.

Lepselter, Susan. *Resonance of Unseen Things*. Bloomington: Indiana University Press, 2015.

Levi-Strauss, Claude. *Tristes Tropiques*. London: Penguin, 2011.

Li, Tania Murray. *Land's End: Capitalist Relations on an Indigenous Frontier*. Durham, NC: Duke University Press, 2014.

Lilienfeld, Amy, and Mette Asmild. "Estimation of Excess Water Use in Irrigated Agriculture: A Data Development Analysis Approach." *Agricultural Water Management* 94, no. 1–3 (2007): 73–82.

Lingley, James R., and Laurence L. Robbins. "Fractures Following Electroshock Therapy." *Radiology* 48, no. 2 (February 1947): 124–28.

Livingston, Julie. *Self-Devouring Growth: A Planetary Parable as Told from Southern Africa*. Durham, NC: Duke University Press, 2019.

Loewen, Royden. *Diaspora in the Countryside: Two Mennonite Communities and Mid-Twentieth Century Rural Disjuncture*. Toronto: University of Toronto Press, 2015.

Longuevergne, Laurent, Bridget Scanlon, and C. R. Wilson. "GRACE Hydrological Estimates for Small Basins: Evaluating Processing Approaches on the High Plains Aquifer, USA." *Water Resources Research* 46 (2010): W11517.

Louderback, David. "Testimony of David Louderback." In US War Department, *Report of the Secretary of War*, 134–41.

Lovewell, J. T. "'Witching' for Water and Other Things." *Transactions of the Kansas Academy of Science* 26 (1913): 103.

MacDonald, A., H. C. Bonsor, K. M. Ahmed, W. G. Burgess, Muhammad Basharat, R. C. Calow, and Ajaya Dixit, et al. "Groundwater Quality and Depletion in the Indo-Gangetic Basin Mapped from In Situ Observations." *Nature Geosciences* 9, no. 10 (October 2016): 762–66.

Maddison, Sarah, Tom Clark, and Ravi de Costa, eds. *The Limits of Settler Colonial Reconciliation: Non-Indigenous People and the Responsibility to Engage.* Singapore: Springer Nature Singapore, 2016.

Marotz, G. A., J. Clark, J. Henry, and R. Standfast. "Cloud Fields over Irrigated Areas in Southwestern Kansas—Data and Speculations." *Professional Geographer* 27, no. 4 (November 1975): 457–61.

Martin, Emily. *Bipolar Expeditions.* Princeton, NJ: Princeton University Press, 2007.

Martinez, Arthur D. "The Historical Development of the Mexican-American Community of Dodge City, Kansas." *International Social Science Review* 58, no. 3 (Summer 1983): 159–67.

Masco, Joseph. *The Future of Fallout, and Other Episodes in Radioactive World-Making.* Durham, NC: Duke University Press, 2021.

———. "The Six Extinctions: Visualizing Planetary Ecological Crisis Today." *Environmental Philosophy* 14, no. 1 (2017): 11–40.

Mayer, Frank H., and Charles B. Roth. *The Buffalo Harvest.* Denver, CO: Sage Books, 1958.

Mays, William. *Sublette Revisited: Stability and Change in a Rural Kansas Community after a Quarter Century.* New York: Florham Park, 1968.

McCabe, Richard, Henry M. Reeves, and Bart W. O'Gara. *Prairie Ghost: Pronghorn and Human Interaction in Early America.* Boulder: University Press of Colorado, 2010.

McConnell, Grant. *Decline of Agrarian Democracy.* New York: Scribner, 1969.

McCormick, Peter J. "The 1992 Secession Movement in Southwest Kansas." *Great Plains Quarterly* 15, no. 4 (Fall 1995): 247–58.

McGraw, Seamus. *A Thirsty Land: The Making of an American Water Crisis.* Austin: University of Texas Press, 2018.

McKittrick, Meredith. "Making Rain, Making Maps: Competing Geographies of Water and Power in Southwestern Africa." *Journal of African History* 58, no. 2 (2017): 187.

———. "Theories of 'Reprecipitation' and Climate Change in the Settler Colonial World." *History of Meteorology* 8 (2017): 74–94.

McKnight, Tom L. "Great Circles on the Great Plains: The Changing Geometry of American Agriculture." *Erdkunde* 33, no. 1 (March 1979): 70–79.

McLaughlin, Thad G. "Accelerated Channel Erosion in the Cimarron Valley in Southwestern Kansas." *Journal of Geology* 55, no. 2 (March 1947): 76–93.

Mead, James R. *Hunting and Trading on the Great Plains, 1859–1875.* Wichita: Rowfant, 2008.

Meadows, William C. *Kiowa Ethnogeography.* Austin: University of Texas Press, 2010.

Mehta, Lyla. "Contexts and Constructions of Water Scarcity." *Economic and Political Weekly* 38, no. 48 (December 2003): 5066–72.

———. *The Politics and Poetics of Water: The Naturalisation of Scarcity in Western India.* Hyderabad: Orient Longman, 2005.

Miller, Henry. *On Writing.* New York: New Directions, 1964.

Minderhoud, P.S.J., G. Erkens, V. H. Pham, Vuong Tran Bui, Laura Erban, Henk Kooi, and E. Stouthamer. "Impacts of 25 Years of Groundwater Extraction on Subsidence in the Mekong Delta, Vietnam." *Environmental Research Letters* 12, no. 6 (2017): 064006.

Miner, Craig. *Next Year Country: Dust to Dust in Western Kansas, 1890–1940.* Lawrence: University Press of Kansas, 2006.

Mishra, Gouri, and Sonia Yeh. "Life Cycle Water Consumption and Withdrawal Requirements of Ethanol from Corn Grain and Residues." *Environmental Science and Technology* 45, no. 10 (May 2011): 4563–69.

Monnett, John H. "Reimagining Transitional Kansas Landscapes: Environment and Violence." *Kansas History* 34 (Winter 2012): 258–79.

Montgomery, M., F. Kamel, T. M. Saldana, M.C.R. Alavanja, and D. P. Sandler. "Incident Diabetes and Pesticide Exposure among Licensed Pesticide Applicators: Agricultural Health Study, 1993–2003." *American Journal of Epidemiology* 167, no. 10 (May 15, 2008): 1235–46.

Mooar, J. Wright. "The First Buffalo Hunting in the Panhandle." *West Texas Historical Association Yearbook* 6: 109–10. Lubbock, TX: West Texas Historical Association, 1930.

Mooar, J. Wright, and Robert F. Pace. *Buffalo Days: Stories from J. Wright Mooar as Told to James Winford Hunt.* Abilene, TX: State House, 2005.

Mooney, James. *Calendar History of the Kiowa Indians.* Washington, DC: Smithsonian Institution Press, 1979.

Moore, Jason. *Capitalism in the Web of Life: Ecology and the Accumulation of Capital.* London: Verso Books, 2015.

Moran-Thomas, Amy. *Traveling with Sugar.* Berkeley: University of California Press, 2019.

Muehlebach, Andrea. "The Price of Austerity: Vital Politics and the Struggle for Public Water in Southern Italy." *Anthropology Today* 33, no. 5 (October 2017): 20–23.

Muehlmann, Shaylih. *Where the River Ends: Contested Indigeneity in the Mexican Colorado Delta*. Durham, NC: Duke University Press, 2013.

Musick, J. T., and F. R. Lamm. "Preplant Irrigation in the Central and Southern High Plains: A Review." *Transactions of the American Society of Agricultural Engineers* 33, no. 6 (November 1990): 1835–42.

Nading, Alex M. "Living in a Toxic World." *Annual Review of Anthropology* 49 (2020): 209–24.

National Agricultural Statistics Service. *Kansas*. Vol. 1, *Complete Report, All Tables, 2017 Census of Agriculture*. Washington, DC: US Department of Agriculture, 2017.

National Beef. "Leaders in Beef Processing." We Know Beef, accessed May 7, 2019. https://www.nationalbeef.com/about-us/we-know-beef/processing-leadership.

National Park Service. "Life of Silas Soule." History & Culture, People, November 14, 2019. https://www.nps.gov/sand/learn/historyculture/the-life-of-silas-soule.htm.

Natural Resources Conservation Service. "KS652.0408 State Supplement: Water Requirements." In *Irrigation Guide, National Engineering Handbook Part 652*, 4.1–4.25. Washington, DC: Natural Resources Conservation Service, 2014.

Neh, Akenji, Andrew Ako Ako, Akoachere Richard Ayuk II, and Takahiro Hosono. "DRASTIC-GIS Model for Assessing Vulnerability to Pollution of the Phreatic Aquiferous Formations in Douala–Cameroon." *Journal of African Earth Sciences* 102 (February 2015): 180–90.

Nelson, R. L. "Assessing Local Planning to Control Groundwater Depletion: California as a Microcosm of Global Issues." *Water Resources Research* 48 (2012): W01502.

Nie, Wanshu, Benjamin F. Zaitchik, Matthew Rodell, Sujay V. Kumar, Martha C. Anderson, and Christopher Hain. "Groundwater Withdrawals under Drought: Reconciling GRACE and Land Surface Models in the United States High Plains Aquifer." *Water Resources Research* 54, no. 8 (2018): 5282–99.

Nilles, Dave. "Conestoga Energy Partners, Fueling the New Frontier." *Ethanol Producer Magazine*, accessed May 7, 2019. http//www.ethanolproducer.com/articles/2482/conestoga-energy-%91fueling-the-new-frontier%27/.

Nixon, Rob. *Slow Violence and the Environmentalism of the Poor*. Cambridge, MA: Harvard University Press, 2011.

Nora, Pierre, and Lawrence D. Kritzman, eds. *Realms of Memory: The Construction of the French Past*. Vol. 1. Translated by Arthur Goldhammer. New York: Columbia University Press, 1996.

North, James B. *Union in Truth: An Interpretive History of the Restoration Movement.* Boston, MA: Standard, 1994.

Northwest Kansas Groundwater Management District 4. "Request for a District-Wide LEMA Submitted to the Chief Engineer, Kansas Department of Agriculture, Division of Water Resources." Kansas Department of Agriculture, June 8, 2017.

———. "Testimony of the Northwest Kansas Groundwater Management District No. 4 (GMD 4) Re: Written Testimony for Proposed District-Wide LEMA of August 23, 2017." Kansas Department of Agriculture, August 28, 2017.

O'Brien, D. M., F. R. Lamm, L. R. Stone, and D. H. Rogers. "Corn Yields and Profitability for Low-Capacity Irrigation Systems." *Applied Engineering in Agriculture* 17, no. 3 (2001): 315–21.

O'Connor, Claire, and Lara Bryant. "Covering Crops: How Federal Crop Insurance Program Reforms Can Reduce Costs, Empower Farmers, and Protect Natural Resources." Issue Papers, Natural Resources Defense Council, December 19, 2017.

O'Connor, Howard Grant, and Thomas J. McClain. *Ogallala Aquifer Study in Kansas: Geohydrology.* Lawrence: University of Kansas, Kansas Geological Survey, 1982.

Ogden, Laura. *Swamplife: People, Gators and Mangroves Entangled in the Everglades.* Minneapolis: University of Minnesota Press, 2011.

Opie, John. *Ogallala: Water for a Dry Land.* 3rd ed. Lincoln: University of Nebraska Press, 2018.

Oppenheimer, Robert. "Acculturation or Assimilation: Mexican Immigrants in Kansas, 1900 to World War II." *Western Historical Quarterly* 16, no. 4 (October 1985): 429–48.

Orlove, Ben, and Steven C. Caton. "Water Sustainability: Anthropological Approaches and Prospects." *Annual Review of Anthropology* 39, no. 1 (October 2010): 401–15.

Orpen, Adela. *Memories of Old Emigrant Days in Kansas, 1861–1865.* London: William Blackwood, 1926.

Orwell, George. *Politics and the English Language.* London: Penguin Classics, 2013.

Ossian, Lisa L. "Bandits, Mad Men, and Suicides: Fear, Anger, and Death in a Troubled Iowa Landscape, 1929–1933." *Agricultural History* 80, no. 3 (July 2006): 296–311.

Ostrom, Elinor. *Governing the Commons: The Evolution of Institutions for Collective Action.* Cambridge: Cambridge University Press, 2015.

Ostrom, Elinor, Thomas Dietz, Nives Dolšak, Paul C. Stern, Susan Stonich, and Elke U. Weber, eds. *The Drama of the Commons.* Washington, DC: National Academies Press, 2002.

Ostrom, Elinor, and Roy Gardner. "Coping with Asymmetries in the Commons: Self-Governing Irrigation Systems Can Work." *Journal of Economic Perspectives* 7, no. 4 (Autumn 1993): 93–112.

Ostrom, Elinor, Paul C. Stern, and Thomas Dietz. "Water Rights in the Commons." *Water Resources Impact* 5, no. 2 (2003): 9–12.

Pandian, Anand, and Stuart J. McLean, eds. *Crumpled Paper Boat: Experiments in Ethnographic Writing.* Durham, NC: Duke University Press, 2017.

Parker, Laura. "What You Need to Know about the World's Water Wars." *National Geographic,* July 14, 2016.

Patterson-Black, Sheryll. "Women Homesteaders on the Great Plains Frontier." *Frontiers: A Journal of Women Studies* 1, no. 2 (Spring 1976): 67–88.

Peck, John C. "Groundwater Management in Kansas: A Brief History and Assessment." *Kansas Journal of Law and Public Policy* 15, no. 441 (November 2006): 441–66.

———. "Kansas Groundwater Management Districts." *University of Kansas Law Review* 29, no. 1 (1980): 51–92.

Perramond, Eric. *Unsettled Waters: Rights, Law, and Identity in the American West.* Berkeley: University of California Press, 2018.

Peterson, Steven M., Amanda T. Flynn, and Jonathan P. Traylor. *Groundwater-Flow Model of the Northern High Plains Aquifer in Colorado, Kansas, Nebraska, South Dakota, and Wyoming.* US Geological Survey Scientific Investigations Report 2016–5153. Reston, VA: US Geological Survey, 2016.

Petryna, Adriana. *Life Exposed.* Princeton, NJ: Princeton University Press, 2003.

———. "What Is a Horizon? Navigating Thresholds in Climate Change Uncertainty." In *Modes of Uncertainty: Anthropological Cases,* edited by Limor Samimian-Darash and Paul Rabinow, 147–64. Chicago: University of Chicago Press, 2015.

———. "Wildfires at the Edges of Science: Horizoning Work amid Runaway Change." *Cultural Anthropology* 33, no. 4 (2018): 570–95.

Pisani, Donald. *From the Family Farm to Agribusiness.* Berkeley: University of California Press, 1984.

Potter, David M. *People of Plenty: Economic Abundance and the American Character.* Chicago: University of Chicago Press, 2009.

Prasch, Thomas. "'Facing This Vast Hardness': The Plains Landscape and the People Shaped by It in Recent Kansas/Plains Film." *Kansas History: A Journal of the Central Plains* 38 (Summer 2015): 108–35.

Provencher, Bill, and Oscar Burt. "A Private Property Rights Regime for the Commons: The Case for Groundwater." *American Journal of Agricultural Economics* 76, no. 4 (November 1994): 875–88.

Purdy, Jedediah. *This Land Is Our Land.* Princeton, NJ: Princeton University Press, 2019.

Ralph, Laurence. *The Torture Letters: Reckoning with Police Violence*. Chicago: University of Chicago Press, 2020.

Ramsey, Michael K. "Kansas Groundwater Management Districts: A Lawyer's Perspective." *Kansas Journal of Law and Public Policy* 15 (2005): 517–24.

Rifkin, Mark. *Beyond Settler Time: Temporal Sovereignty and Indigenous Self-Determination*. Durham, NC: Duke University Press, 2017.

Riney-Kehrberg, Pamela. *Rooted in Dust: Surviving Drought and Depression in Southwestern Kansas*. Lawrence: University Press of Kansas, 1994.

Roberts, Elizabeth. "What Gets Inside: Violent Entanglements and Toxic Boundaries in Mexico City." *Cultural Anthropology* 32, no. 4 (2017): 592–619.

Roberts, Liz, and Katherine Phillips, eds. *Water, Creativity and Meaning: Multidisciplinary Understandings of Human-Water Relationships*. New York: Routledge, 2018.

Roberts, Martha G., Timothy D. Male, and Theodore P. Toombs. *Potential Impacts of Biofuels Expansion on Natural Resources: A Case Study of the Ogallala Aquifer Region*. New York: Environmental Defense Fund, 2007.

Roberts, R. L. "Expansion of Church of Christ in West Texas, 1870–1900." *West Texas Historical Association Year Book* 53 (1977).

Rochín, Refugio I. "Introduction: Latinos on the Great Plains: An Overview." *Great Plains Research* 10, no. 2 (Fall 2000): 243–52.

Rodell, Matthew, and James S. Famiglietti. "The Potential for Satellite-Based Monitoring of Groundwater Storage Changes Using GRACE: The High Plains Aquifer, Central US." *Journal of Hydrology* 263, no. 1–4 (2002): 245–56.

Rodell, Matthew, Isabella Velicogna, and James S. Famiglietti. "Satellite-Based Estimates of Groundwater Depletion in India." *Nature* 460, no. 7258 (August 20, 2009): 999–1002.

Rodriguez, V. M., Mona Thiruchelvam, and Deborah A. Cory-Slechta. "Sustained Exposure to the Widely Used Herbicide Atrazine: Altered Function and Loss of Neurons in Brain Monoamine Systems." *Environmental Health Perspectives* 113, no. 6 (June 2005): 708–19.

Roediger, David R. *The Wages of Whiteness: Race and the Making of the American Working Class*. London: Verso, 1999.

Rogers, Danny H. *Efficiencies and Water Losses of Irrigation Systems*. Manhattan, KS: Kansas State University Cooperative Extension Services, 1997.

Rogers, Danny H., and Freddie R. Lamm. "Soil Water Survey after Corn Harvest in Northwest Kansas." *Applied Engineering in Agriculture* 10, no. 1 (1994): 37–40.

Roitman, Janet. *Anti-crisis*. Durham, NC: Duke University Press, 2013.

Rojas, Rodrigo, Luc Feyen, and Alain Dassargues. "Conceptual Model Uncertainty in Groundwater Modeling: Combining Generalized Likelihood Uncertainty Estimation and Bayesian Model Averaging." *Water Resources Research* 44, no. 12 (2008): W12148.

Sachs, Carolyn E. *Gendered Fields: Rural Women, Agriculture and Environment.* New York: Routledge, 2018.

Sagala, J. K., and Z. A. Smith. "Comparative Groundwater Management: Findings from an Exploratory Global Survey." *Water International* 33 (2008): 258–67.

Sanderson, Matthew R., and R. Scott Frey. "From Desert to Breadbasket . . . to Desert Again? A Metabolic Rift in the High Plains Aquifer." *Journal of Political Ecology* 21 (2014): 516–32.

———. "Structural Impediments to Sustainable Groundwater Management in the High Plains Aquifer of Western Kansas." *Agriculture and Human Values* 32, no. 3 (September 1, 2015): 401–17.

Santa Fe Trail Association. "Awards of Merit." *Wagon Tracks* 6, no. 1 (November 1991): 1, 3–4.

Scanlon, Bridget R., Claudia C. Faunt, Laurent Longuevergne, Robert C. Reedy, William M. Alley, Virginia L. McGuire, and Peter B. McMahon. "Groundwater Depletion and Sustainability of Irrigation in the US High Plains and Central Valley." *Proceedings of the National Academy of Sciences of the USA* 109, no. 24 (June 12, 2012): 9320–25.

Scanlon, Bridget R., Kelley E. Keese, Alan L. Flint, Lorraine E. Flint, Cheikh B. Gaye, W. Michael Edmunds, and Ian Simmers. "Global Synthesis of Groundwater Recharge in Semiarid and Arid Regions." *Hydrological Processes* 20, no. 15 (2006): 3335–70.

Schumm, Stanley Alfred, and Robert W. Lichty. *Channel Widening and Flood-Plain Construction along Cimarron River in Southwestern Kansas: Erosion and Sedimentation in a Semiarid Environment.* Washington, DC: Government Printing Office, 1963.

Schwieder, Dorothy, and Deborah Fink. "Plains Women: Rural Life in the 1930s." *Great Plains Quarterly* 8, no. 2 (Spring 1988): 79–88.

Scranton, Roy. *Learning to Die in the Anthropocene.* San Francisco: City Lights Books, 2015.

Seaboard Foods. "Facts and Figures Processing." About, accessed May 7, 2019. https://seaboardfoods.com/About/Pages/Facts_Figures_Plant.aspx.

Sears, Louis, Joseph Caparelli, Clouse Lee, Devon Pan, Gillian Strandberg, Linh Vuu, and C.-Y. Cynthia Lin Lawell. "Jevons' Paradox and Efficient Irrigation Technology." *Sustainability* 10, no. 5 (May 2018): 1590.

Sebald, W. G. *The Natural History of Destruction.* New York: Random House, 2003.

———. *Rings of Saturn.* New York: New Directions, 1998.

———. *Vertigo.* New York: New Directions, 2000.

Selfa, Theresa, Albert Iaroi, and Morey Burnham. "Promoting Ethanol in Rural Kansas: Local Framings and Cultural Politics." *Journal of Rural Studies* 39 (June 1, 2015): 63–73.

Seton, Ernest T. *Life-Histories of Northern Animals: An Account of the Mammals of Manitoba*. New York: Scribner, 1909.

Sexson, Mark, and Jerry Choate. "Historical Biogeography of the Pronghorn in Kansas." *Transactions of the Kansas Academy of Science* 84, no. 3 (1981): 128–33.

Shah, Tushaar, Aditi Deb Roy, Asad S. Qureshi, and Jinxia Wang. "Sustaining Asia's Groundwater Boom: An Overview of Issues and Evidence." *Natural Resources Forum* 27, no. 2 (May 2003): 130–41.

Shahid, Shamsuddin, Xian-June Wang, M. Moshiur Rahman, Rashidul Hasan, Sobri Bin Harun, and Supiah Shamsudin. "Spatial Assessment of Groundwater Over-Exploitation in Northwestern Districts of Bangladesh." *Journal of the Geological Society of India* 85, no. 4 (2015): 463–70.

Shankar, P. S. Vijay, Himanshu Kulkarni, and Sunderrajan Krishnan. "India's Groundwater Challenge and the Way Forward." *Economic and Political Weekly* 46, no. 2 (January 2011): 37–45.

Shapin, Steven. *A Social History of Truth*. Chicago: University of Chicago Press, 1994.

Shen, Hong, Marc Leblanc, Sarah Tweed, and Wenzhao Liu. "Groundwater Depletion in the Hai River Basin, China, from In Situ and GRACE Observations." *Hydrological Sciences Journal* 60, no. 4 (2015): 671–87.

Sherow, James. *The Grasslands of the United States: An Environmental History*. Santa Barbara, CA: ABC-CLIO, 2007.

———. *Watering the Valley: Development along the High Plains Arkansas River, 1870–1950*. Lawrence: University Press of Kansas, 1991.

Shi, Jiansheng, Zhao Wang, Zhaoji Zhang, Yuhong Fei, Yasong Li, Jingsheng Chen, and Yong Qian. "Assessment of Deep Groundwater Over-Exploitation in the North China Plain." *Geoscience Frontiers* 2, no. 4 (2011): 593–98.

Shumsky, Neil L. "Dust, Disease, Death and Deity: Constructing and Deconstructing the 'Dust Bowl.'" *Journal of American Culture* 38, no. 3 (September 2015): 218–31.

Sittig, Ann L., and Martha Florinda González. *The Mayans among Us: Migrant Women and Meatpacking on the Great Plains*. Lincoln: University of Nebraska Press, 2016.

Slotkin, T. A. "Does Early-Life Exposure to Organophosphate Insecticides Lead to Prediabetes and Obesity?" *Reproductive Toxicology* 31, no. 3 (April 2011): 297–301.

Smarsh, Sarah. *Heartland: A Memoir of Working Hard and Being Broke in the Richest Country on Earth*. New York: Scribner, 2018.

Smith, Michael M. "Beyond the Borderlands: Mexican Labor in the Central Plains, 1900–1930." *Great Plains Quarterly* 1, no. 4 (Fall 1981): 239–51.

Smith, Zadie. "Elegy for a Country's Seasons." *New York Review of Books* 61, no. 6 (April 3, 2014). https://www.nybooks.com/articles/2014/04/03/elegy-countrys-seasons/.

Smits, David D. "The Frontier Army and the Destruction of the Buffalo: 1865–1883." *Western Historical Quarterly* 25, no. 3 (Autumn 1994): 312–38.

Snyder, N. D. "Testimony of N.D. Snyder." In US War Department, *Report of the Secretary of War*, 76–81.

Soll, David. *Empire of Water: An Environmental and Political History of the New York City Water Supply*. Ithaca, NY: Cornell University Press, 2013.

Sophocleous, Marios. "From Safe Yield to Sustainable Development of Water Resources: The Kansas Experience." *Journal of Hydrology* 235 (2000): 27–43.

——— "Groundwater Management Practices, Challenges and Innovations in the High Plains Aquifer, USA: Lessons and Recommended Actions." *Hydrogeology Journal* 18, no. 3 (2010): 559–75.

Soule, Silas S. "Testimony of Captain Silas S. Soule." In US War Department, *Report of the Secretary of War*, 8–19.

Southwest Kansas Groundwater Management District Number 3. "Revised (2018) (Draft) Management Program." Accessed March 1, 2019. http://www.gmd3.org.

———. *Revised Management Program*. Draft June 2020. Garden City, KS: Southwest Kansas Groundwater Management District Number 3. http://www.gmd3.org/wp-contents/uploads/2020/07/Management-Program-June-Draft.pdf.

Splinter, William E. "Center-Pivot Irrigation." *Scientific American* 234, no. 6 (June 1976): 90–99.

Steedman, Carolyn. *Landscape for a Good Woman*. New Brunswick, NJ: Rutgers University Press, 1986.

Steinberg, Ted. *Nature Incorporated: Industrialization and the Waters of New England*. Cambridge: Cambridge University Press, 1991.

Stern, Paul C., Thomas Dietz, and Elinor Ostrom. "Research on the Commons: Lessons for Environmental Resource Managers." *Environmental Practice* 4, no. 2 (2002): 61–64.

Stewart, Kathleen. *Ordinary Affects*. Durham, NC: Duke University Press, 2006.

———. *A Space on the Side of the Road: Cultural Poetics in an Other America*. Princeton, NJ: Princeton University Press, 1996.

Steyl, G., and I. Dennis. "Review of Coastal-Area Aquifers in Africa." *Hydrogeology Journal* 18, no. 1 (February 2010): 217–25.

Stoler, Ann Laura. *Carnal Knowledge and Imperial Power: Race and the Intimate in Colonial Rule*. Berkeley: University of California Press, 2010.

———. *Imperial Debris: On Ruins and Ruination*. Durham, NC: Duke University Press, 2013.

———. "'Interior Frontiers' as Political Concept, Diagnostic, and Dispositif." In *The Rise of Trumpism*, Hot Spots series edited by Lucas Bessire and David Bond, Society for Cultural Anthropology, January 18, 2017. https://culanth.org/fieldsights/interior-frontiers-as-political-concept-diagnostic-and-dispositif.

Stone, L. R., A. J. Schlegel, F. R. Lamm, and W. E. Spurgeon. "Storage Efficiency of Preplant Irrigation." *Journal of Soil and Water Conservation* 49, no. 1 (January 1994): 72–76.

Strassberg, Gil, Bridget R. Scanlon, and Don Chambers. "Evaluation of Groundwater Storage Monitoring with the GRACE Satellite: Case Study of the High Plains Aquifer, Central United States." *Water Resources Research* 45, no. 5 (2009): W05410.

Stull, Donald D. "I Come to the Garden: Changing Ethnic Relations In Garden City, Kansas." *Urban Anthropology and Studies of Cultural Systems and World Economic Development* 19, no. 4 (Winter 1990): 303–20.

Stull, Donald D., Michael J. Broadway, and David Griffith. *Any Way You Cut It: Meat Processing and Small-Town America.* Lawrence: University Press of Kansas, 1995.

Suter, Margaret. "An Update on Yemen's Water Crisis and the Weaponization of Water." *MENASource* (blog), Atlantic Council, November 29, 2018. https://www .atlanticcouncil.org/blogs/menasource/an-update-on-yemen-s-water-crisis-and -the-weaponization-of-water/.

Svobida, Lawrence. *Farming the Dust Bowl: A First Hand Account from Kansas.* Lawrence: University Press of Kansas, 1986.

Swaminathan, Krishnan. "Pesticides and Human Diabetes: A Link Worth Exploring?" *Diabetic Medicine* 30, no. 11 (November 2013): 1268–71.

Swanson, Nancy, Andre Leu, Jon Abrahamson, and Bradley Wallet. "Genetically Engineered Crops, Glyphosate and the Deterioration of Health in the United States of America." *Journal of Organic Systems* 9, no. 2 (January 2014): 6–37.

Sylvester, Kenneth M., and Paul W. Rhode. "Making Green Revolutions: Kansas Farms, Recovery, and the New Agriculture, 1918–1981." *Agricultural History* 91, no. 3 (2017): 342–68.

Taussig, Michael. *The Corn Wolf.* Chicago: University of Chicago Press, 2014.

———. "The Corn-Wolf: Writing Apotropaic Texts." *Critical Inquiry* 37, no. 1 (Autumn 2010): 26–33.

———. *Palma Africana.* Chicago: University of Chicago Press, 2018.

———. *Shamanism, Colonialism and the Wild Man.* Chicago: University of Chicago Press, 1987.

Taylor, Jonathan G., Thomas R. Stewart, and Mary Downton. "Perceptions of Drought in the Ogallala Aquifer Region." *Environment and Behavior* 20, no. 2 (1988): 150–75.

Tiernes, Brock. "Groundwater Citizenship and Water Supply Awareness: Investigating Water-Related Infrastructure and Well Ownership." *Rural Sociology* 83, no. 2 (2018): 347–75.

Torell, L. Allen, James D. Libbin, and Michael D. Miller. "The Market Value of Water in the Ogallala Aquifer." *Land Economics* 66, no. 2 (May 1990): 163–75.

Travis, Paul D. "Charlatans, Sharpers, and Climatology: The Symbolism and Mythology of Late Nineteenth-Century Expansionism in Kansas." PhD dissertation, University of Oklahoma, 1975.

Tsing, Anna L. *Mushroom at the End of the World*. Princeton, NJ: Princeton University Press, 2015.

Tsing, Anna L., Nils Bubandt, Elaine Gan, and Heather Anne Swanson, eds. *Arts of Living on a Damaged Planet: Ghosts and Monsters of the Anthropocene*. Minneapolis: University of Minnesota Press, 2017.

Tuck, Eve, and K. Wayne Yang. "Decolonization Is Not a Metaphor." *Decolonization: Indigeneity, Education and Society* 1, no. 1 (2012): 1–40.

Turner, Frederick Jackson. *The Frontier in American History*. New York: Holt, 1921.

Tyson Foods. *Investor Fact Book: Fiscal Year 2017*. Springdale, AR: Tyson Foods, 2017.

USDA Economic Research Service. "Tailored Reports." Agricultural Resource Management Survey (ARMS), accessed June 7, 2019. https://my.data.ers.usda.gov/arms/tailored-reports.

USDA National Agricultural Statistics Service (NASS). *2012 Census of Agriculture County Profile: Grant County, Kansas*. Washington, DC: USDA-NASS, 2018. www.nass.usda.gov/Publications/AgCensus/2012/Online_Resources/County_Profiles/Kansas/cp20067.pdf.

———. *2012 Census of Agriculture County Profile: Haskell County, Kansas*. Washington, DC: USDA-NASS, 2018. www.nass.usda.gov/Publications/AgCensus/2012/Online_Resources/County_Profiles/Kansas/cp20081.pdf.

USDA Risk Management Agency. *Corn: Kansas*. Fact sheet, 2018 crop year. Topeka, KS: Risk Management Agency Topeka Regional Office, 2018.

———. "Crop Indemnity Maps Archive." May 7, 2019. https://legacy.rma.usda.gov/data/indemnity/archive.html.

US War Department. *Report of the Secretary of War, Communicating, in Compliance with a Resolution of the Senate of February 4, 1867, a Copy of the Evidence Taken at Denver and Fort Lyon, Colorado Territory, by a Military Commission, Ordered to Inquire into the Sand Creek Massacre, November, 1864*. Washington, DC: Government Printing Office, 1867.

Valdés, Dennis Nodín. "Settlers, Sojourners, and Proletarians: Social Formation in the Great Plains Sugar Beet Industry, 1890–1940." *Great Plains Quarterly* 10, no. 2 (Spring 1990): 110–23.

VanLooy, Jeffrey, and Charles Martin. "Channel and Vegetation Change on the Cimarron River, Southwestern Kansas, 1953–2001." *Annals of the Association of American Geographers* 95, no. 4 (December 1, 2005): 727–39.

Van Steenbergen, Frank, Allah Bakhsh Kaisarani, Niamat Ullah Khan, and Mohammed Shamshad Gohar. "A Case of Groundwater Depletion in Balochistan, Pakistan: Enter into the Void." *Journal of Hydrology* 4, pt. A (September 2015): 36–47.

Velmurugan, Ganesan, Tharmarajan Ramprasath, Mithieux Gilles, Krishnan Swaminathan, and Subbiah Ramasamy. "Gut Microbiota, Endocrine-Disrupting Chemicals, and the Diabetes Epidemic." *Trends in Endocrinology and Metabolism* 28, no. 8 (August 2017): 612–25.

Vogeler, Ingolf. *The Myth of the Family Farm: Agribusiness Dominance of US Agriculture.* New York: Routledge, 2019.

Vogt, Evon, and Peggy Golde. "Some Aspects of the Folklore of Water Witching in the United States." *Journal of American Folklore* 71, no. 282 (October–December 1958): 519–31.

Vogt, Evon, and Ray Hyman. *Water Witching, U.S.A.* Chicago: University of Chicago Press, 1959.

Von Schnitzler, Antina. *Democracy's Infrastructure: Techno-politics and Protest after Apartheid.* Princeton, NJ: Princeton University Press, 2016.

Wada, Yoshihide, Min-Hui Lo, Pat J.-F. Yeh, John T. Reager, James S. Famiglietti, Ren-Jie Wu, and Yu-Heng Tseng. "Fate of Water Pumped from Underground and Contributions to Sea-Level Rise." *Nature Climate Change* 6, no. 8 (August 2016): 777–80.

Wada, Yoshihide, Ludovicus P. H. van Beek, and Marc F. P. Bierkens. "Nonsustainable Groundwater Sustaining Irrigation: A Global Assessment." *Water Resources Research* 48 (2012): W00L06.

Wada, Yoshihide, Ludovicus P. H. van Beek, Cheryl M. van Kempen, Josef W.T.M. Reckman, Slavek Vasak, and Marc F. P. Bierkens. "Global Depletion of Groundwater Resources." *Geophysical Research Letters* 37, no. 20 (2010): L20402.

Wada, Yoshihide, Ludovicus P. H. van Beek, F.C.S. Weiland, B. F. Chao, Y. H. Wu, and Marc F. P. Bierkens. "Past and Future Contribution of Global Groundwater Depletion to Sea-Level Rise." *Geophysical Research Letters* 39, no. 9 (2012): L09402.

Wagner, John. "Water and the Commons Imaginary." *Current Anthropology* 53, no. 5 (October 2012): 617–41.

Wahl, Ana-Maria, Steven E. Gunkel, and Thomas W. Sanchez. "Death and Disability in the Heartland: Corporate (Mis)Conduct, Regulatory Responses, and the Plight of Latino Workers in the Meatpacking Industry." *Great Plains Research* 10, no. 2 (October 2000): 329–57.

Walker, T. S., R. P. Singh, and M. Asokan. "Risk Benefits, Crop Insurance, and Dryland Agriculture." *Economic and Political Weekly* 21, no. 25/26 (June 21–28, 1986): A81–88.

Walley, Christine. *Exit Zero.* Chicago: University of Chicago Press, 2013.

Warren, Carol. *Madwives: Schizophrenic Women in the 1950s.* New Brunswick, NJ: Rutgers University Press, 1987.

Waters, John D. "The Water Supply of the Farmhouse." *Bulletin of the Kansas State Agricultural College* 42, no. 45 (March 1916): 15.

Webb, Walter Prescott. *The Great Plains*. Boston, MA: Ginn, 1931.

Weber, Jeremy G., Nigel Key, and Erik O'Donoghue. "Does Federal Crop Insurance Make Environmental Externalities from Agriculture Worse?" *Journal of the Association of Environmental and Resource Economists* 3, no. 3 (September 2016): 707–42.

Whitacre, Christine, and Stephen DeVore. "Cultural Resource Investigations at the Lower Cimarron (Wagon Bed) Spring Campsite." *Kansas Anthropologist* 19 (1998): 7–35.

White, Richard. *The Organic Machine: The Remaking of the Columbia River*. New York: Hill and Wang, 1995.

White, Stephen E. "Migration Trends in the Kansas Ogallala Region and the Internal Colonial Dependency Model." *Rural Sociology* 63 (1998): 253–71.

Whittemore, Donald, James Butler Jr., and B. B. Wilson. "Assessing the Major Drivers of Water-Level Declines: New Insights into the Future of Heavily Stressed Aquifers." *Hydrological Sciences Journal* 61 (2016): 134–45.

Whittemore, Donald, Brownie Wilson, and James Butler Jr. *Status of the High Plains Aquifer in Kansas*. Technical Series. Lawrence: Kansas Geological Survey, September 2018.

Wislizenus, Frederick Adolph. *Memoir of a Tour to Northern Mexico, Connected with Col. Doniphan's Expedition, in 1846 and 1847*. Washington, DC: Tippin and Streeper, 1848.

Wolfe, Patrick. "Settler Colonialism and the Elimination of the Native." *Journal of Genocide Research* 8, no. 4 (2006): 387–409.

Woo, Sang-Woo, and Sang-Hwan Do. "Tongue Laceration during Electroconvulsive Therapy." *Korean Journal of Anesthesiology* 62, no. 1 (January 2012): 101–2.

Woods, John, and Brownie Wilson. *The Water Information Storage and Retrieval Database (WISARD, a.k.a. WIZARD) Web Site User Manual*. Open File Report. Lawrence: Kansas Geological Survey, 2007.

Worster, Donald. *Dust Bowl: The Southern Plains in the 1930s*. 25th anniversary ed. New York: Oxford University Press, 2004.

———. *Rivers of Empire: Water, Aridity and the Growth of the American West*. London: Oxford University Press, 1992.

Wright, Robert Marr. *Dodge City: The Cowboy Capital*. Wichita: Wichita Eagle, 1913.

Wu, JunJie, and Christian Langpap. "The Price and Welfare Effects of Biofuel Mandates and Subsidies." *Environmental and Resource Economics* 62, no. 1 (2015): 35–57.

Wuthnow, Robert. *Red State Religion: Faith and Politics in America's Heartland*. Princeton, NJ: Princeton University Press, 2011.

Wutich, Amber. "Water Insecurity: An Agenda for Research and Call to Action for Human Biology." *American Journal of Human Biology* 32, no. 1 (2020): e23345.

Wutich, Amber, and Melissa Beresford. "The Economic Anthropology of Water." *Economic Anthropology* 6, no. 2 (June 2019): 168–82.

ILLUSTRATIONS

Maps

The Ogallala Aquifer, 2015. (By Nat Case)   viii

The Little Rock House. (By Nat Case)   ix

Figures

1. Irrigation motor, 2018.   3

2. Satellite image of southwest Kansas fields. (Google Earth)   6

3. Pivot sprinkler, 2019.   15

4. Water in the Cimarron, 1910. (Photograph courtesy of Historic Adobe Museum, Ulysses, Kansas)   19

5. Drilling a water well, 1910. (Photograph courtesy of Historic Adobe Museum)   23

6. At Wagonbed Springs, 2018.   25

7. Water drive.   34

8. Flood irrigation, 2018.   41

9. Sunday rabbit drive in Grant County, ca. 1940. (Photograph courtesy of Historic Adobe Museum)   51

10. Bison skulls by railroad tracks in Dodge City, 1870s. (Photograph courtesy of Boot Hill Museum, Inc., Dodge City, Kansas)   54

Photographs by author unless otherwise stated.

11. Cloudbank over Cimarron.   63

12. Aerial image of Wagonbed Springs, 1939. (Photograph from Fern Bessire archives)   71

13. Blow sand and dry field, 2018.   81

14. Breaking out prairie and listing land on contour in Grant County, 1936. (Photograph courtesy of Historic Adobe Museum)   85

15. Dust storm near Beaver, Oklahoma, April 14, 1935. (Photograph courtesy of National Archives and Records Administration)   87

16. Early irrigation well in Haskell County, 1941. (Photograph courtesy of National Archives and Records Administration)   90

17. Wheat harvest crew, 1930. (Photograph courtesy of Historic Adobe Museum)   107

18. Marginalia.   117

19. Carbon plant, 2019.   122

20. Digging an irrigation ditch along the Arkansas River, 1909. (Photograph courtesy of Historic Adobe Museum)   125

21. Scraper.   130

22. Nozzle.   137

23. Reflections in the North Fork of the Cimarron in Grant County, 1928. (Photograph courtesy of Historic Adobe Museum)   144

24. Mexican-American laborers in an irrigated onion field in Grant County, 1945. (Photograph courtesy of Historic Adobe Museum)   147

25. At the Little Rock House, 1947. (Photograph from Fern Bessire archives)   153

26. Evening clouds.   154

27. Farmhand.   163

INDEX

Page numbers in italics refer to figures.

Abilene Christian University, 93, 95, 100

Abraham, Negev Plain, 155

Adobe Walls, 103

Agent Orange, 122

agri-biopolitics, 206n15

agribusiness, 4, 17, 132, 156, 168; chaos of, 168; clearing the Plains for, 131; corporations, 31, 79–80, 112; economics of, 39, 58, 110; logics, 9; myths, 6–7; pesticides, 31; writing, 189n11

Agricultural Adjustment Administration (AAA) program, 87–88

alfalfa, 61, 76–78, 132

American Restoration Movement, 198n1

antelope, 18, 51, 55–56

anthropology/anthropologists, 8–9, 16, 84, 120, 181, 207nn19–20

aqueducts, 15, 109–110, 199n6

aquifer(s), xiii–xiv; alfalfa and, 77–78; decline of, 2–4, 15, 110, 139, 141, 149, 157, 188n8, 191n39; drying, 180–181; Dust Bowl and, 88–89; future of, 137–138; governance, 11–15, 56–57; groundwater extraction, 3; High Plains system, xi, 4, 5, 7, 111; irrigation and, 74; layers, 7, 133; loss, 5, 7, 9, 16, 47, 60, 109–110, 133, 135, 168–169, 175, 176, 179, 188n8; minerals, 34; Ogallala, xi, 4, 140; replacing water in, 109–110; use of, 80, 112, 130; waters, 8, 10, 17, 20, 24, 40, 78, 112–114, 164, 170

Arabian peninsula, 3, 179

Argentina, 148, 177, 179

Arkansas River, 48, 114, 124, 135; digging irrigation ditch along, 125; valley, 47, 138, 200n13

atrazine, 123–124, 200n11

Australia, 3, 177

Autry, Gene, 90

Baldwin, James, responsibility, 205n5

Bangladesh, 177

basketball, 99–100, 102

Bear Creek, 45, 125

Beatty Ranch, 64

Beaver, Oklahoma, dust storm, 87

Beck Plan, 110

Bell, Earl, 84–86, 88–90

Bent, George, 127, 131

Bessire, Fern (grandmother), 21–25, 182; "A Long Wait for Harvest" (story), 114–116, 117; blue binder, 116–117, 133–134; brother Gene, 29, 93–94, 97–105, 147; buffalo skulls, 54, 54; clock, 119; Dust Bowl and, 82–83; inspiration by, 171; interview of Pearl, 36–37; irrigation file, 22–23, 24; marginalia, 117, 169; in mental hospital, 68–70, 73; nervous breakdowns, 21, 67–68; radio program, 22, 83, 131; shock therapy, 69, 73. See also RW (great-grandfather)

Big Circle, 28, 29, 63, 163

bison, 18; buried bones, 1, 46; extermination of, 47, 130, 193n13; final, in Grant County, 50–51; guns for killing, 48–49; herds, 48, 162, 195n36; hunters, 18; Inman's calculation of, 196n52; lacing carcasses with strychnine, 51; skulls of, 54, 167. See also buffalo

black blizzards, 86

Black Kettle, 126, 127

blow sand and dry field (2018), 81

Brookover, Earl, 24

Brown, H. H., 24, 52

buffalo, accounts of last, 195n36; bones, 1, 17, 45–47, 53, 129; extermination of, 47, 130, 193n13; hides, 48, 196n46; hunters/hunting, 50, 52, 53, 127, 143, 195n36; killing, 46–50, 52, 53, 55, 175; metal silhouette of bull, 25, 25, 182; skulls, 54, 54. See also bison

Buffalo Bones, 53

Buffalo Bull River, 50

Buffalo City, 47–48

buffalo grass, 25, 34, 63

Buffalo Wallow, 103

Buffalo War, 50, 129

buffalo wild, 48

California, 3, 102, 177

Camp Amache, 125

capitalism/capitalist, 88, 89, 168, 205n2

carbon plants, 122; bones to, 53; carbon black, 121, 199–200n9

Carter, Floyd, on settler surveys, 195n39

Catholics, 95

chemical spray, 34, 63, 169

Chile, 3, 177

China, 3, 177, 178

Chivington (town), 125

Chivington Canal Company, 125, 132

Chivington, John, as Colorado Volunteer, 126–128, 200n15

Christianity, 94

Church of Christ, 93–95, 198n1

Cibolero bison hunters, 18

Cimarron River, 1, 17, 36–37, 63, 99, 105, 114, 130, 132, 161, 179; bison and, 50–51, 55, 143; cloudbank over, 63; description of, 18–20; groundwater stream, 17; High Plains aquifer and, 190n31; irrigation and, 124; Lower Springs of, 18, 70; metal buffalo, 25, 25, 182; nineteenth-century surveyor on, 18–19; North Fork of, in Grant County (1928), 144; photograph of water in (1910), 19; wells on, 33

clouds, 153–155, 154

Colorado Volunteer militias, 126, 128

Columbian Carbon Company, 122

Conservation Reserve Enhancement Program (CREP), 75–76, 197n67, 203n10

conservative(s), 157; in Christianity,
    94; groundwater use, 5, 139;
    political groups, 112; rural, 171
Cook, James, 53
Cook, John, on hunting, 49–50, 52
corn, 1–2, 42, 44, 57, 100, 124, 136;
    cornfields, 2, 29, 45, 61; ethanol
    plants, 78, 192n2; flood irrigation,
    41, 45; irrigation, 39, 77, 192n1;
    prewatering, 192n46; prices, 30–31,
    40–41, 156–157, 160
Corn Management (Kansas State
    University), 40, 192n1
COVID-19 pandemic, 203–204n17
coyotes, 20–21, 101, 164–165, 182
crop insurance, 42–43, 193nn8, 10, 12
Cropquest, 57

dairy, 61, 76–77, 112
Darton, Nelson H., on subterranean
    formations, 191n42
Dead Line, 50
Deep Earth, feminine power, 131
defoliant 2,4,5-T, Agent Orange,
    122–123, 200n10
democracy, xiv, 5, 172–173
diabetes, 124, 200n12
Dillon, Karen, 78
dioxin, 122–123, 200n10
Dirty Thirties, 82, 83
dispossession, 46, 129, 131, 179, 181
Division of Water Resources (DWR),
    59, 62, 138, 157, 183, 190n30
Dixon, Billy, 50, 52
Dobritzhoffer, Martin, missionary,
    189n17
doctrine of prior appropriation, 156
Dodge, Richard Irving, Lieutenant
    Colonel, 48–50, 53–54

Dodge City, 52, 53, 54, 149
drilling water well (1910), 23
Dr. Parker's Painless Panacea, 91
Dust Bowl, 5, 19; analysis of, 197n77;
    black blizzards, 86; corporations
    and, 79; dust storm (1935), 87; Fern
    and, 82–83; Haskell County and,
    85–86, 88–89; homesteaders in, 160;
    irrigation farmers in, 148
dust cloud, 37, 91, 94

ecologies, 5, 82, 85, 168, 199n6
Edwards, A. D., on drought, 85–86, 88
electroconvulsive therapy (ECT), 73
electroshock therapy, 69, 73,
    196–197n64
Elliot, T. S., 9
environmental crisis, 10
Environmental Protection Agency, 58,
    123, 204n17
Espy, James, on rainmaking, 189n17
ethanol plants, 40, 78, 112, 173, 192n2
Expanded Premium Services, 58

family farms, 40, 79–80, 100, 191n44
Farm Bill, section 1999A, 58
Farm Bureau, 158
Farmhand, 162, 163
farming, 65, 90, 100; corn, 39;
    corporate hog, 203n15; dryland, 98,
    108; Dwane (farmer), 135–139, 158,
    170; family, 40, 79–80, 100, 191n44;
    government payments in, 88;
    groundwater and, 5; industrial, 78,
    86, 88; insurance, 42–43; irrigation,
    13, 28, 52, 130, 148, 156, 167; no-till,
    123–124; "Robert" on, 28–32; role of
    women in, 24, 191n44; sustainable,
    89; wheat harvest crew (1930), 107

Federal Crop Insurance Corporation, 193n12

federal government: Agricultural Adjustment Administration program (AAA), 87–88; Conservation Reserve Program, 203n10; drought emergency, 61, 87; farm insurance, 42–43; investigations, 128; Wagon-bed Springs as historic landmark, 134; wars against Native peoples, 47; Works Progress Administration projects, 87

Federal Land Bank, 20

FieldScout moisture probe, 58

film reels (16 mm), 105–106

Finney county, 138

Fischer, Otto, 24

flood irrigation, 24, *41*, 45, 103

Fort Hays State University, 46, 195n39

Fort Lyon, 126, 200n15

Friesen v. Barfield, 202n8

Future Farmers of America (FFA), 137, 138

Garden City, 149

Garden City Coop, 40

Garetson Brothers v. American Warrior, Inc., 204nn21–22

genocides, 46, 129, 132, 178, 194n34

Glimpses of Grant County (radio program), 83

global pandemic, 168, 203–204n17

global water crisis, 2

glyphosate, 162

Granada, 47, 53, 55

Gran Chaco of Paraguay, 8, 9, 64–65, 189n17

Grandmother, power, 131

Grant County, 17, 24, 50, 83, 89; breaking out prairie and listing land on contour, *85*; cumulative payments, 193n12; estimated recharge, 188n6; farm census (2012), 193n11; Mexican-American laborers in irrigated onion field, *147*; North Fork of the Cimarron in, *144*; Sunday rabbit drive (ca.1940), *51*

Gregg, Josiah, merchant diary, 18, 143

groundwater: analyses of, 187n1; aquifer declines, 2–4; dependence on, 4–5; economic value for Kansas, 108–109; global water crisis, 2–3; management, 16, 59–60, 111, 140, 149, 172, 174, 179, 202nn4–5, 204n20, 205n3; sheet water by Humphreys, 23–24

Groundwater Management District (GMD): Northwest, 140–141. *See also* Southwest Groundwater Management District (GMD)

Harvest America Corporation, 203n17

Haskell County, 61, 84–87, 89, 188n6, 193nn11–12; Agricultural Adjustment Administration (AAA) payments to, 87–88; cumulative payments, 193n12; estimated recharge, 188n6; farm census (2012), 193n11; irrigation well, *90*

herbicides, 39, 77, 162, 200n10

hermeneutic violence, 205n2

High Plains aquifer, 4, 103, 199n7; Cimarron River and, 190n31; depletion of, 4–5; description of, 7–8; irrigated lands in, 199n2; losses of, 177, 188n8; overuse of, xi; variation within, 111

Hudson Valley, 65

Humphreys, William, 23
hydrogeology, 205n6
hydrologists, 4, 157, 188n4
hydrology, 57, 158, 205n6
Hydroresource, 58
hyperproductivity, 31

India (northern), 3, 177, 205n3
Indian(s): Arapaho, 126; bones and
    relics, 37; Cheyenne, 17, 50, 126–128,
    130, 131, 179; Comanche, 17, 50, 128,
    130, 179; Kiowa, 17, 50, 128, 130, 179;
    Pawnee, 114; Plains Apache, 17, 179
Indian Territory, 17, 49, 128; Indian
    Caves, 37; Indian stories, 22, 130–131
individualism, 88, 169
Informing People on Pesticides
    program, 204n17
Inman, Henry, on bison, 196n52
insanity, 50, 53–54, 170
intensive groundwater unit control
    areas (IGUCAs), 138, 174
irrigation, 2, 97–99, 135; alfalfa and corn,
    77; aquifer and, xi, 191n39, 199n2;
    buried gas lines for motors, 74; center
    pivot system, 29; Chivington Canal,
    132; deep well, 5–6; ditch along
    Arkansas River (1909), 125; farming,
    13, 28–30, 130, 141, 148, 155–156, 167,
    174; Fern's folder on, 22–24; flood,
    24, 41, 45, 103; groundwater table
    and, 89; Haskell County well (1941),
    90; limited insurance, 193n8; motors,
    2, 3, 14, 20, 39, 105, 133, 164, 182; nozzle,
    80, 135, 137; rise of, 74, 89; South
    Fork Irrigation Corporation, 124;
    as stabilizer of agriculture, 89; tech-
    nological advances in, 80, 136–137,
    168, 177, 202n2; total value from,

109; Wagonbed Springs, 74; water
    rights for, 91; western Kansas, 114, 139
Israel, 177

jackrabbit roundups, 51, 51
Jones, Charles J. ("Buffalo"), on
    buffalo decline, 195n36
Jordan, 177

Kansas: agriculture in southwest, 6;
    Division of Water Resources, 157;
    economic value of groundwater,
    108–109; Grant County, 17, 24, 50,
    83, 85, 89, 144, 147; Groundwater
    Management Districts (GMDs),
    11–15; Haskell County, 61, 84–87, 89,
    90, 188n6, 193nn11–12; Northwest
    GMD, 140–141; satellite image of
    fields, 6; State Board of Agriculture,
    156, 158. See also Southwest Ground-
    water Management Districts
    (GMDs)
Kansas City Journal, 112
Kansas Corn, 57, 192nn2–3
Kansas Geological Survey (KGS), 20,
    57, 111, 138, 199n7
Kansas Leadership Council, 112
Kansas State University (KSU), 138;
    Corn Management guide, 40, 192n1;
    extension officers, 136; Water
    Technology Farms, 135
Kearney county, 138
Kearney-Finney LEMA (KFL)
    Steering Committee, 138–139, 141
Kelman, Ari, 129
knob knocking, 103

lambda-cyhalothrin, 162
Lauer, Stephen, 140

litigation, plaintiff (Jay) in aquifer case, 155–161

Little Rock House, 1, 9, 11–12, 27–29, 32, 34, 65, 85, 121, 135, 145, 156; agribusiness in county, 80; antelope and, 55–56; aquifer decline, 2, 188n8; aquifer recharge rates near, 14; aquifer under, 160; barn at, 45; bison and, 47, 50; Cimarron River, 17–20, 130; Dust Bowl era, 82; farms near, 43; Fern and, 21–24, 82, 152, 162, 166; groundwater monitoring, 4, 20, 139; Lower Springs of Cimarron, 18; map of, ix; photograph of RW and son (1947), 153; purchase of, 93; Susana and, 146–147, 150; well at, 2

local enhanced management area (LEMA), 138–139, 141, 174, 203n9

Lower Springs National Historic Landmark, 134

McCormick combine (1948), 106
McRae, Billy, 51
marginalia, 117, 169
massacres, 94, 127–129, 130, 130; Herrero and San, 179; Indigenous peoples, 47; Sand Creek, 129, 132, 179
Mayer, Frank, 48–49, 51–52
Mays, William, 89
meat-packing plants, 27, 56, 78, 113, 148–149, 167, 203n15, 203n17
Medicine Lodge Treaty (1867), 50
megadairies, 76–77, 112
Mehne spirits, 131
Mencken, H. L., 84
Mennonites, 120, 148–149, 160, 189n17
mental health crisis, 145
metabolic rift, 199n6
meter tampering, 42

Mexican American laborers, 101, 147, 148, 149
Missouri River, 109–110
Molner, Joseph, on electroshock, 197n64
Murzuk-Djado Basin, 177

Namibia, 177, 179
National Historic Landmark, 134
National Park Service, 129, 134, 201n21
Negev Plain, 155
Nepal, 177
New Mexico, 18, 20
Nixon, Tom, 48
No Man's Land, 50
North China Plain, 3
Northwest Groundwater Management District (GMD), 140–141. See also Southwest Groundwater Management District (GMD)

Ogallala aquifer, xi, 4, 202n2; agricultural economics, 198–199n2; efforts to slow decline, 173–174; fate of, 179; groundwater loss, 4, 140–141, 160, 188n8; label of hydrogeological layer, 191n42; map of, viii
Old Neese Place, 21, 97
Old Testament, 155
organochlorines, 124
organophosphates, 124

Pakistan, 177, 178
Palo Duro country, 103
Pearl (settler), interview of, 36–37
pesticides, 30–31, 39, 123–124
pivot irrigation system, 29–30, 136, 165; sand hills and, 75; sprinkler, 15; Water Tech Farm, 135

Plains insanity, 53–54

The Plow that Broke the Plains (newsreel), 90

pollution, 2, 199–200n9

population, 206n15; of beef cattle, 197n70; of county in Kansas, 56, 85, 89, 148; growth, 2; herbicides and resistant species, 162

prewatering, 32, 136, 192n46

Profiler Plus, 58

racism, 149

radio program, Fern's, 22, 83, 131

Rath City, 52

Rattlesnake Creek, 48

ravens, 51

Rea, Charles, 24

responsibility, xiii–xiv, 10, 190n20, 205n5; blame and, 96, 175; depletion and, 176; embracing mutual, 161; for future, 164; grit as, 31–32; motherhood and marriage, 115; sense of, 174; taking, xiii–xiv, 10, 105; taking care of land as moral, 160

Rio Grande Valley, 18

Rocky Mountain News (newspaper), 127–128

Rusinow, Irving, 84

RW (great-grandfather), 5, 17, 19–20, 99–101, 106; church and, 67–68, 95; death of, 93, 100; Farmhand, 162, 163, *163*; irrigation system and, 29; successor, 32. *See also* Bessire, Fern (grandmother)

Sand Creek, 125, 128, 130–131, 132; Battle Ground, 126; massacre, 129, 179; national historic site, 132

Sanderson, Matthew, 140

Santa Fe Railroad, 47, 49, 109

Santa Fe Trail, 18, 21

Santa Fe Trail Association, 134

satellite image, Kansas field, 6

Scott, Buell, 24

scraper, 64, 65, *130*

Sebald, W. G., writings of, 204–205n1

Servitech, 58

settler colonialism, 168, 193n14

Sharps Buffalo rifles, 48–49

sheet water, Humphreys, 23–24

Smith, Mr. and Mrs., 26–27

soil moisture sensors, 135, 136

sorghum, 1, 78, 124, 192n2

Soule, Silas, on Indians, 128, 201n21

South Africa, 177

South Fork Irrigation Corporation, 124

Southwest Groundwater Management District (GMD), 11–15, 56, 59, 112; board, 11, 157, 159; board meeting, 107–114; draft management plan, 190nn22–23; groundwater depletion, 88, 107; LEMA creation and, 138–139; management document of, 111; membership, 11; mission of, 108–109; Northwest GMD and, 140–141; policy of controlled decline, 140; standards for new wells, 13

spray plane, 162

Staked Plains, 50

Stone-Campbell Movement, 198n1

strychnine, 51

suicide, 86, 173, 203n13

Syria, 177

Tamil Nadu, 178

Taussig, Michael, 189n11, 205n2

Technology Minutes, 57

tetrachlorodibenzodioxin, 122–123

Texas beef, 49, 194n24
Third Colorado Cavalry Regiment, 126, 129
Towler, Lucille, 191n36
truth squads, 84
Tumbling Tumbleweeds (film), 90
Turkey, 177

uranium, 48, 58, 114
US Army Corps of Engineers, 110
US Department of Agriculture: Agricultural Resource Management Survey, 192n3; Bureau of Agricultural Economics, 85; Risk Management Agency (USDA-RMA), 183, 193n12, 193n8

Valley Model 1060 sprinkler, 29
Volga German families, 148
Vollmer factory (Germany), 49

Wadi Fatima, Saudi Arabia, 178
Wagonbed Springs, 18, 20–21, 24, 69, 125, 179, 182; aerial image of (1939), 71; Dixon's winter hide camp, 50; irrigation and, 24, 74; National Historic Landmark, 134; photograph (2018), 25
War Bonnet, 201n21

War Department, 128
Washita River, 128
water: management, 16, 59–60, 111, 140, 149, 172, 174, 179, 202n5, 204n20; Valley Model 1060, 29, 34, 34; well drilling (1910), 23; water conservation areas (WCAs), 15–16, 138; water witching, 7–8, 189n15. *See also* aquifer(s); Division of Water Resources (DWR); groundwater
Water Boot Camp, 138
Water Technology Farms, 135
Wellington, 94
Wells, Betty, 71–72
What Kind of Child Were You? (quiz), 117–118
wheat harvest crew (1930), 107
White Antelope, 127, 201n21
White supremacy, 9, 168
Wilson, R. R., 24
WIZARD (open-access database), 110, 199n7
wolves, 18, 51, 195n39
Woofter v. McClaskey, 202n8
Wynkoop, Ned, on Indians, 126, 128, 201n21

Yellow House Canyon, 103
Yemen, 177, 178